APPLICATIONS OF PLANT CELL AND TISSUE CULTURE

The Ciba Foundation is an international scientific and educational charity. It was established in 1947 by the Swiss chemical and pharmaceutical company of CIBA Limited—now CIBA-GEIGY Limited. The Foundation operates independently in London under English trust law.

The Ciba Foundation exists to promote international cooperation in biological, medical and chemical research. It organizes about eight international multidisciplinary symposia each year on topics that seem ready for discussion by a small group of research workers. The papers and discussions are published in the Ciba Foundation symposium series. The Foundation also holds many shorter meetings (not published), organized by the Foundation itself or by outside scientific organizations. The staff always welcome suggestions for future meetings.

The Foundation's house at 41 Portland Place, London, W1N 4BN, provides facilities for meetings of all kinds. Its Media Resource Service supplies information to journalists on all scientific and technological topics. The library, open seven days a week to any graduate in science or medicine, also provides information on scientific meetings throughout the world and answers general enquiries on biomedical and chemical subjects. Scientists from any part of the world may stay in the house during working visits to London.

APPLICATIONS OF PLANT CELL AND TISSUE CULTURE

A Wiley – Interscience Publication

1988

JOHN WILEY & SONS

Chichester · New York · Brisbane · Toronto · Singapore

© Ciba Foundation 1988

Published in 1988 by John Wiley & Sons Ltd, Baffins Lane,
Chichester, Sussex PO19 1UD, UK.

Suggested series entry for library catalogues:
Ciba Foundation Symposia

Ciba Foundation Symposium 137
× + 269 pages, 33 figures, 16 tables

Library of Congress Cataloging-in-Publication Data

Applications of plant cell and tissue culture.
    p. cm. — (Ciba Foundation symposium ; 137)
  "A Wiley–interscience publication."
  Includes indexes.
  ISBN 0 471 91886 5
1. Plant propagation — In vitro — Congresses.   2. Plant
biotechnology — Congresses.   3. Plant cell culture — Congresses.
4. Plant tissue culture — Congresses.   I. Series.
SB123.6.A67   1988
631.5′23 — dc 19   88-10687
                                                                                     CIP

British Library Cataloguing in Publication Data

Applications of plant and cell tissue culture.
  — (Ciba Foundation symposium ; 137).
  1. Plants. Cells & tissues. Cultures
  I. Series
  581′.0724

  ISBN 0 471 91886 5

Typeset by Inforum Ltd, Portsmouth
Printed and bound in Great Britain by the Bath Press Ltd., Bath, Avon

Contents

Participants

Professor Dr W. Barz Lehrstuhl für Biochemie der Pflanzen, Westfälische Wilhelms-Universität, Hindenburgplatz 55, D-4400 Münster, Federal Republic of Germany

Dr R.S. Chaleff American Cyanamid Co, PO Box 400, Princeton, New Jersey 08540, USA

Professor E.C. Cocking Plant Genetic Manipulation Group, Department of Botany, School of Biological Sciences, University of Nottingham, University Park, Nottingham NG7 2RD, UK

Professor J. Durzan Department of Environmental Horticulture, College of Agriculture & Environmental Sciences, Agricultural Experimental Station, Davis, California 95616, USA

Professor M.W. Fowler Wolfson Institute of Biotechnology, University of Sheffield, Sheffield S10 2TN, UK

Dr Y. Fujita Bioscience Research Laboratories, Mitsui Petrochemical Industries Ltd, 6-1-2 Waki-cho, Kuga-Gun, Yamaguchi-Ken 740, Japan

Professor Esra Galun Department of Plant Genetics, The Weizmann Institute of Science, PO Box 26, Rehovot 76100, Israel

Professor T.C. Hall Department of Biology, Texas A & M University, College Station, Texas 77843-3258, USA

Professor H. Harada Institute of Biological Sciences, University of Tsukuba, Sakura-Mura, Nihari-Gun, Ibaraki-Ken 305, Japan

Dr C.T. Harms Biotechnology Research, CIBA-GEIGY Corporation, PO Box 12257, Research Triangle Park, North Carolina 27709-2257, USA

Dr A. Hirai Graduate Division of Biochemical Regulation, Faculty of Agriculture, Nagoya University, Furo-cho, Chikusa, Nagoya 464, Japan

Professor A. Komamine   Department of Plant Physiology, Biological
Institute, Tohoku University, Aoba-Yama, Sendai 980, Japan

Mrs M.V. Mhatre (*Ciba Foundation Bursar*)   Plant Biotechnology Section,
Bio-organic Division, Bhabha Atomic Research Centre, Trombay,
Bombay 400085, India

Dr K. Ohyama   Research Center for Cell & Tissue Culture, Faculty of
Agriculture, Kyoto University, Yoshida, Sakyo-ku, Kyoto 606, Japan

Professor Dr I. Potrykus   Institut fur Pflanzenwissenschaften, Eidgenossische
Technische Hochschule Zentrum LFV-E20, CH-8092 Zurich, Switzerland

Dr M.J.C. Rhodes   Plant Cell Biotechnology Group, AFRC Institute of Food
Research, Norwich Laboratory, Colney Lane, Norwich NR4 7UA, UK

Sir Ralph Riley   16 Gog Magog Way, Stapleford, Cambridge CB2 5BQ, UK

Dr W.R. Scowcroft   Vice President, Research and Development, Biotechnica
Canada Inc, 170 6815-8 Street NE, Calgary, Alberta, Canada T2E 7H7

Professor M. Sugiura   Center for Gene Research, Nagoya University,
Furocho, Chikusa-ku, Nagoya 464, Japan

Professor M. Tabata   Department of Pharmacognosy, Faculty of
Pharmaceutical Sciences, Kyoto University, Yoshida, Shimoadachi-cho,
Sakyo-ku, Kyoto 606, Japan

Professor I. Takebe   Department of Biology, Faculty of Science, Nagoya
University, Furo-cho, Chikusa, Nagoya 464, Japan

Dr H. Uchimiya   Institute of Biological Sciences, University of Tsukuba,
Sakura-Mara, Ibaraki-Ken 305, Japan

Professor M. van Montagu   Laboratorium Genetika, Rijksuniversiteit
Gent, Ledeganckstraat 35, B-9000 Gent, Belgium

Dr L.A. Withers[*]   Department of Agriculture & Horticulture, University of
Nottingham, School of Agriculture, Sutton Bonington,
Loughborough LE12 5RD, UK

---

[*] *Present address*: IBPGR Headquarters, Food and Agriculture Organization of the United
Nations, Via delle Terme di Caracalla, 00100 Rome, Italy

Professor Y. Yamada (*Chairman*)   Research Center for Cell & Tissue Culture, Faculty of Agriculture, Kyoto University, Yoshida, Sakyo-ku, Kyoto 606, Japan

Professor M.H. Zenk   Lehrstuhl Pharmazeutische Biologie, Universitat Munchen, Karlstrasse 29, D-8000 Munchen 2, Federal Republic of Germany

Introduction

People have used plants as sources of materials other than nutrients for almost as long as they have been eating the plants themselves. The use of plant-derived medicines, poisons and narcotics is still common in many cultures, and herbal medicines in particular are becoming increasingly popular in the Western world. Against this, there is the continuing desire within our scientifically oriented society to improve on the achievements of Nature. In agriculture, this was most obvious in the development of synthetic herbicides and pesticides, and the extensive application of artificial fertilizers to produce ever greater yields. These farming methods were complemented by breeding programmes, initiated when the first plant species were chosen for deliberate cultivation, and now designed to select plants with increased yield potential.

The ability of 'weeds and pests' to develop resistance to man-made products, particularly when they were applied in large amounts and were extremely stable so that they persisted in the environment, and concern over their safety when incorporated into human and animal food chains, made these procedures less desirable, on both social and economic grounds. The emphasis in science turned to finding natural means of resistance and transferring these to commercially important crops. Similarly, the ability of certain plants, the legumes, in association with rhizobacteria, to fix their own nitrogen, thereby circumventing the need for the major exogenous fertilizer, encouraged the hope that this trait could be introduced into other crop species. The time scale of conventional plant breeding programmes is slow and genetic engineering of plant cells seemed to be the ideal way of introducing specific desirable characteristics into the relevant plant. However, this approach encountered two major problems. Isolation of the desired gene was hindered by the fact that the biochemical or genetic basis of the phenotype was unknown, so that the time required for screening procedures was determined by the generation time of the plant. In addition, manipulation of plant genotypes by the techniques of molecular biology proved to be much more difficult than similar work on animal cells. This was partially due to the presence of the outer plant cell wall, a barrier which scientists had to overcome when trying to introduce genetic material into a plant cell. The fact that some traits were encoded by the organellar genomes made genetic analysis more complicated.

Plant cell biology has now reached a stage when most, if not all, of these problems can be resolved. Much of this progress is due to the development of both cell and tissue culture technologies. Regeneration of whole plants from single cells has been possible since the 1930s but for a long time it was restricted to a few isolated species. This has now been extended to include the major commercial crops – both food plants, such as the cereals, and cash crops like tobacco. An important breakthrough has been the establishment of protoplast cultures (plant cells stripped of their outer wall), which greatly facilitates genetic manipulation of the cells. The protoplasts can then be stimulated to reform a cell wall and develop into whole plants. A direct commercial application of plant cell or tissue cultures is in the area of secondary metabolite production. Many compounds of use, particularly in the pharmaceutical or food-processing industries, are produced in plant cells as by-products of normal metabolic pathways, and either excreted or, more usually, stored in intracellular vacuoles. Extraction of these products from whole plants is an expensive process; if the cells from the relevant plant tissue can be grown in culture under conditions such that they produce high levels of these compounds, then purification procedures are greatly simplified. At present, only two such chemicals are produced on an industrial scale in this way, both in Japan, as described by Dr Fujita, but several other processes are being developed. The papers presented in this book describe various aspects of plant cell and tissue culture, especially the ways in which these may be used to exploit the natural bounty of products made by plants.

The Editors

Herbicide-resistant plants from cultured cells

R. S. Chaleff

Department of Central Research and Development, Experimental Station, E.I. Du Pont and Company, Wilmington, Delaware 19898, USA

Abstract. Tobacco mutants resistant to sulphonylurea herbicides were isolated by selection in cell culture. Resistance resulted from mutations in either of two unlinked genes encoding isozymes of the branched chain amino acid biosynthetic enzyme, acetolactate synthase (ALS). These mutant alleles directed the synthesis of forms of ALS with greatly reduced sensitivity to inhibition by sulphonylurea herbicides. The cloning and transfer of these mutant alleles encoding herbicide-insensitive forms of ALS will provide an efficient and powerful method by which to introduce herbicide resistance into presently susceptible crop species.

1988 Applications of plant cell and tissue culture. Wiley, Chichester (Ciba Foundation Symposium 137) p 3–20

By bestowing upon higher plants many of the experimental advantages that were previously restricted to microorganisms, cell and protoplast culture provide a powerful and efficient method by which to modify plants genetically. *In vitro* culture permits large populations of physiologically and developmentally homogeneous plant cells to be produced and maintained in a nutritionally and chemically controlled environment. That environment can then be altered to inhibit the growth of normal cells and thereby establish growth conditions that are selective for defined mutant types. Enzymic digestion of the cell wall to form protoplasts makes high frequencies of genetic transformation possible by removing a barrier to the uptake of foreign DNA.

Despite these technical developments, so far only a small number of transformed and mutant plants have been produced through cell culture and very few of the modifications that have been introduced are of agronomic interest. Several limitations of cell culture systems are responsible for this disappointingly modest success. First, plant regeneration has not yet been achieved from cultured cells and/or protoplasts of many crop species. Second, most agronomically important traits are genetically complex (polygenic) and may not be altered qualitatively by mutation at a single locus. Third, only

---

* *Present address*: American Cyanamid Company, PO Box 400, Princeton, New Jersey 08540, USA.

modifications of traits expressed by cultured cells and not of traits that are exclusively functions of the whole plant can be selected *in vitro* (Chaleff 1983). Fourth, our fragmentary knowledge of the molecular and cellular bases of many agronomically important traits prevents the design of effective mutant selection strategies or the cloning and transfer of genes that underlie such traits.

In view of these considerations, herbicide resistance appears as a trait of potential agronomic interest that is better suited than most such traits to modification *in vitro*. Herbicides that interfere with basic metabolic functions can be expected to inhibit growth of cultured cells. In such cases, resistance can be selected directly and unambiguously by growth in the presence of a normally inhibitory concentration of the herbicide. Moreover, resistance can often be expected to be dominant to sensitivity and to result from alteration of only a single gene. These latter features enable the application of molecular technologies, such as molecular cloning, *in vitro* mutagenesis and genetic transformation, to the introduction of herbicide resistance.

In vitro selection for resistance to sulphonylurea herbicides

Tobacco (*Nicotiana tabacum* cv. Xanthi) mutants resistant to the two sulphonylurea herbicides, chlorsulphuron and sulphometuron methyl (the active ingredients in Glean® and Oust®, respectively), were selected by transferring callus cultures initiated from leaves of haploid plants to medium supplemented with the appropriate herbicide at a concentration of 6 nM (Chaleff & Ray 1984). In some experiments, genetic variability of the cultures was increased prior to selection by treatment with the chemical mutagen ethylnitrosourea. Extensive crosses with regenerated diploid plants established that resistance of 10 isolates resulted from single semidominant nuclear mutations. Two genetic loci, designated *SuRA* and *SuRB*, were identified by linkage studies performed with six of the mutants. Mutations at both loci, although selected *in vitro*, confer increased resistance on the whole plant (Chaleff & Bascomb 1987; Fig. 1).

Biochemical characterization of the resistant mutants led to identification of acetolactate synthase (ALS), the first enzyme specific to the isoleucine-leucine-valine biosynthetic pathway, as the site of action of chlorsulphuron and sulphometuron methyl (Chaleff & Mauvais 1984). Extracts of suspension cultures of cells homozygous for a mutation (*S4*) at the *SuRB* locus contained a form of ALS that was far less sensitive to inhibition by these two sulphonylurea herbicides than was the corresponding activity in normal cell extracts. Conclusive proof that the altered ALS activity was the basis for resistance and therefore, that ALS was the site of action of the herbicide, was provided by the demonstration of co-segregation through genetic crosses of the herbicide-insensitive ALS activity with the resistant phenotype. Resistant ALS activity

FIG. 1. Normal (left) and homozygous *S4/S4* mutant (right) tobacco plants in the field without herbicide application (background) and following treatment with sulphometuron methyl at 30 g/acre (courtesy of B Smeeton and K Bridle, RJ Reynolds Tobacco Co).

is also present in leaves of both *SuRA* and *SuRB* mutant plants (Chaleff & Bascomb 1987; Fig. 2). The non-hyperbolic inactivation of ALS activities in extracts of both *SuRA* and *SuRB* mutants to a plateau of approximately 50% of the initial activity suggests the presence of two forms of the enzyme. Because these activities are encoded by separate genes, but their non-mutant forms possess the same catalytic activity and are sensitive to inhibition by the same compounds, they can be regarded as ALS isozymes, although attempts to resolve them chromatographically have not yet been successful (Chaleff & Bascomb 1987).

Passage of homozygous *S4/S4* mutant callus tissue through a second cycle of selection in the presence of 600 nM sulphometuron methyl yielded an even more highly resistant cell line. Genetic studies with regenerated plants revealed that this enhanced level of resistance resulted from the occurrence of a second mutation (*Hra*), which was linked to the *S4* mutation and, therefore, resided at or near the *SuRB* locus. Plants homozygous for both mutations (*S4 Hra/S4 Hra*) possessed a high proportion of an ALS activity that was almost insensitive to herbicide inhibition and were at least fivefold more resistant to damage by chlorsulphuron than were plants of the singly mutant *S4/S4* parental genotype (G.L. Creason & R.S. Chaleff, unpublished results).

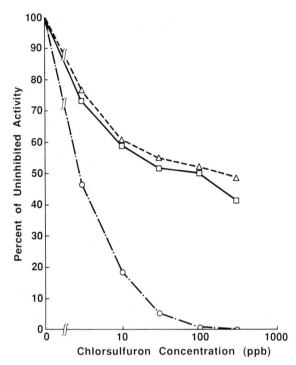

FIG. 2.   Responses to chlorsulphuron of acetolactate synthase activities in leaves of normal ($+/+; +/+$) and resistant mutant plants. Activities are presented as percentages of the respective uninhibited acetolactate synthase activities: O·—·O, normal; △-----△, *S4/S4; +/+*; □——□, *+/+;C3/C3*. ppb, parts per one thousand million.

Molecular cloning of mutant plant acetolactate synthase

Because of the large size of higher plant genomes, it is not a trivial matter to clone a higher plant gene that is not either multiply represented in the genome or abundantly expressed during development, if the product of that gene has not been purified. Fortunately, it proved possible to gain experimental access to the tobacco ALS gene through a simpler microbial system. Cloning of the ALS gene (*ILV2*) from the yeast, *Saccharomyces cerevisiae*, was accomplished by exploiting the increased tolerance for sulphometuron methyl that is expressed by cells containing additional copies of the ALS gene, which direct overproduction of the enzyme (Falco & Dumas 1985). Herbicide-resistant colonies containing increased concentrations of ALS activity were isolated after transformation of sensitive yeast cells with a yeast genomic library in a high copy number plasmid vector. DNA sequence analysis of the cloned yeast gene revealed an unexpectedly high degree of homology between the deduced amino acid sequences of the yeast and *Escherichia coli* enzymes in three distinctly conserved domains (Falco et al

```
                                                                                                      100
Yeast ALS    MIRQSTLKNFAIKRCFQHIAYRNTPAMRSVALAQRFYSSSSRYYSASPLPASKRPEPAPSFNVDPLEQAEPSKLAKKLRAEPDMDTSFVGLTGGQIFNE
                                                                                                      ||
E. coli Isozyme II                                                                               MNGAQWVVH

                                                                                                      200
             MMSRQNVDTVFGYPGGAILPVYDAIHNSDKFNFVLPKHEQSAGHMAEGYARASGKPGVVLVTSGPGATNVVTPMADAFADGIPMVVFTGQVPTSAIGTDA
             ||  ||||||||| ||||    | |||||  |||| |  |||   ||  || | |||| | ||||||||| ||  | ||  |||| ||||| |||||  | ||
             ALRAQGVNTVFGYPGGAIMPVYDAL YDGGVEHLLCRHEQGAAMAAIGYARATGKTGVCIATSGPGATNLITGLADALLDSIPVVAITGQVSAPFIGTEA

                                                                                                      300
             FQEADVVGISRSCTKWNVMVKSVEELPLRINEAFEIATSGRPGPVLVDLPKDVTAAILRNPIPTKTTLPSNALNQLTSRAQDEFVMQSINKAADLINLAK
             || || |||   |||  |   ||| ||||||  |||  ||||||||||| |||        | ||   |          |||     || ||  |||||   |
             FQEVDVLGLSLACTKHSFLVQSLEELPRIMAEAFDVACSGRPGPVLVDIPKDIQLASGDLEPWFTTVENEVTFPHAEVEQARQMLAK

                                                                                                      400
             KPVLYVGAGILNHADGPRLLKELSDRAQIPVTTTLQGLGSFDQEDPKSLDMLGMHGCATANLAVQNADLIIAVGARFDDRVTGNISKFAPEARRAAAEGR
             ||  ||| |  |    |  ||  | ||  ||      |||||   ||   ||  ||||| ||  ||   || | |||||||||||||  |   | |   |||
             KPMLYYGGGVGMAQAVPA LREFLAATKMPATCTLKGLGAVEADYPYYLGMLGMHGTKAANFAVQECDLLIAVGARFDDRVTGKLNTFAPHAS

                                                                                                      500
             GGIIHFEVSPKNINKVVQTQIAVEGDATTNLGKMMSKIFPVKERSEWFAQINKWKKEYPYAYMEETPGSKIKPQTVIKKLSKVANDTGRHVI VTTGVGQ
             ||  ||  |   ||  |  | ||| ||| || |  |   | |   |      |   ||      | |       ||| ||  | |  |       |||  |||  |||
             VIHMDIDPAEMNKLRQAHVALEGDLNALLPALQQPLNQYDMWQQHCAELRDEHSWRYDHPGDAIYAPLLLKQLSDRKPADCV      VTTDVGQ

                                                                                                      600
             HQMWAAQHWTWRNPHTFITSGGLGTMGYGLPAAIGAQVAKPESLVIDIDGDASFNMTLTELSSAVQAGTPVKILILNNEEQGMVTQWQSLFYEHRYSHTH
             ||||||||| |    | |||| |||||| ||||| ||||| | ||  | || |  | |  | || |||  || | | ||||  | ||  |  | | |||
             HQMWAAQHIAHTRPENFITSSGLGTMGFGLPAAVGAQVARPNDTVVCISGDGSFMMNVQELGTVKRKQLPLKIVLLDNQRLGMVRQWQQLFFQERYSET

             QL NPDFIKLAEAMGLKGLRVKKQEELDAKLKEFVSTKGPVLLEVEVDKKVPVLPMVAGGSGLDEFINFDPEVERQQTELRHKRTGKH
             ||    ||||  |  || || | |    |  | |   |  ||   |   |  |   || | |||| |    |||
             LTDNPDFLMLASAFGIHGQHITRKDQVEAALDTMLNSDGPYLLHVSIDELENWPLVPPGASNSEMLEKLS
```

FIG. 3. Comparison of deduced amino acid sequences of yeast and *E. coli* acetolactate synthases. The vertical bars indicate amino acid residues conserved between the two proteins (Falco et al 1985).

```
         10              30              50              70              90             110
GGATCCCTCTTTCATTTGTTCTCATCCATTTTGCGATTCATCATGTGCATTTAATCAGTAGGACCCCTTTTTAGCTTAGTAGTGCTCTCAACTTAATATTAAACCAACCACACT

        130             150             170             190             210             230
CCATCTGCATTACCCTCCTTCCTTCCAGTTTCGTCTCTCCCTGCCCTCCCTTCAACATGGCGGCGGCTCCATCCTCCGTTTCTCCAAAACCCTATCGCCTTCCTCCTCC
                                                     M  A  A  A  P  S  P  S  S  S  A  F  S  K  T  L  S  P  S  S  S

        250             270             290             310             330             350
ACATCTCCACCCTCTCCTCCTAGATCAACCTTCCCTTCCCCACCACCCCCACAAGACACCCCACCTCACCCACCCCTCACCTCCCCCGTGCGCGTCGT
 T  S  S  T  L  L  P  R  S  T  F  P  F  P  H  H  P  H  K  T  T  P  P  P  L  H  T  H  I  H  S  Q  R  R  R

        370             390             410             430             450             470
TTCACCATATCCAATGTCATTTCCACTAACCAAAAAGTTCCAGACCGAAAAAACTTTCGTTTCCGGTTTGCCTGACGAACCCAGAAAGGGTTCCGACGTTCTCCTGGAG
 F  T  I  S  N  V  I  S  T  N  Q  K  V  S  Q  T  E  K  T  E  T  F  V  S  R  F  A  P  D  E  P  R  K  G  S  D  V  L  V  E

        490             510             530             550             570             590
GCTCTCGAAAGACAAGGGGGTACGGACGTCTTTGCGTACCCAGGTGGCGCTTCCATGGAGATTCACCAAGCTTTGACCCGTTCACCGTGCAACGTGCCCACGTCACGACGAG
 A  L  E  R  E  G  V  T  D  V  F  A  Y  P  G  G  A  S  M  E  I  H  Q  A  L  T  R  S  S  I  I  R  N  V  L  P  R  H  E  Q

        610             630             650             670             690             710
GGCGGTGTCTTCGCCGCTGAGGGTTACGCACGCGCCACCGGATTTCCCGGCGTTTGCATTGCCGTTTGCATTGGCCGCCACCAATCGTGCACGGCCGCCTCGCCGACTGGAT
 G  G  V  F  A  A  E  G  Y  A  R  A  T  G  F  P  G  V  C  I  A  T  S  G  P  G  A  T  N  L  V  S  G  L  A  D  A  L  L  D

        730             750             770             790             810             830
AGGTCCCCATTGTTGCTATAAACAGGTCAAGTGCCACGTAGGATGATAGGTACTGCTGTTTTCGAGGAACTCCTATTGTTGAGGTAACTAGATCGATTACCAAGCATAATTATCTCGTT
 S  V  P  I  V  A  I  T  G  Q  V  P  R  R  M  I  G  T  D  A  F  Q  E  T  P  I  V  E  V  T  R  S  I  T  K  H  N  Y  L  V

        850             870             890             910             930             950
ATGGACGTAGAGGATATTCCTAGGGTTGTACGTGAAGCTTTTTTCTCGCGAGATCGGGCCGGCCTGGCCCTATTTGATTGATGTACCTAAGGATATTCAGCAACAATTGGTGATACCT
 M  D  V  E  D  I  P  R  V  V  R  E  A  F  F  L  A  R  S  G  R  P  G  P  I  L  I  D  V  P  K  D  I  Q  Q  L  V  I  P

        970             990            1010            1030            1050            1070
GACTGGGATCAGCCAATGAGGTTACCTGGTTACATGTCTAGGTTGCCTAAATTGCCCAATGAGATGCTTTTAGAACAAATTGTTAGGCTTATTTCTGAGTCAAAGAAGCCTGTTTTGTAT
 D  W  D  Q  P  M  R  L  P  G  Y  M  S  R  L  P  K  L  P  N  E  M  L  L  E  Q  I  V  R  L  I  S  E  S  K  K  P  V  L  Y

       1090            1110            1130            1150            1170            1190
GTGGGGGGTGGGTTGTTCCAGAACCAGTGAGGACTTGAGACGATTCGTGGAGCTCACGGGTATCCCGTGGCAAGTACTTTGATGGGTCTTGGAGCTTTTCCAACTGGGGATGAGCTTTCC
 V  G  G  G  C  S  Q  S  S  E  D  L  R  R  F  V  E  L  T  G  I  P  V  A  S  T  L  M  G  L  G  A  F  P  T  G  D  E  L  S
```

```
        1210        1230        1250        1270        1290        1310
CTTTCAATGTTGGGTATGCATGGTACTGTTTATGCTAATTATGCTGTTGGACAGTAGTGATTTGTTGCTCGCATTGGGGTGAGGTTGATGATAGAGTTACTGGAAAGTTAGAAGCTTTT
 L  S  M  L  G  M  H  G  T  V  Y  Y  A  N  Y  A  V  D  S  S  D  L  L  L  A  F  G  V  R  F  D  D  R  V  T  G  K  L  E  A  F

        1330        1350        1370        1390        1410        1430
GCTAGCCGACAAAAATTGTTCACATTGATTCAGCTGAGATTGAAAGAACAAGCAGCCTAGTTCCATTTGTCAGATATCAAGTTGGCGTTACAGGGTTTGAATTCGATA
 A  S  R  A  K  I  V  H  I  D  I  D  S  A  E  I  G  K  N  K  Q  P  H  V  S  I  C  A  D  I  K  L  A  L  Q  G  L  N  S  I

        1450        1470        1490        1510        1530        1550
CTGGAGAGTAAGGAAGGTAAACTGAAGTGGATTTTTCTGCTTGGAGGCAGGAGTTGACGAGCAGAAAGTGAAGCACCCATTGAACTTTAAAACTTTTGGTGATGCAATTCCTCCGCAA
 L  E  S  K  E  G  K  L  K  L  D  F  S  A  W  R  Q  E  L  T  E  Q  K  V  K  H  P  L  N  F  K  T  F  G  D  A  I  P  P  Q

        1570        1590        1610        1630        1650        1670
TATGCTATCCAGGTTCTAGATGAGTTAACTAATGGGAATGCTATTATAAGTACTGGTGTGGGCAACACCAGATGTGGGCTGCTCAATACTATAAGTACAGAGAAAGCCACGCCAATGGTTG
 Y  A  I  Q  V  L  D  E  L  T  N  G  N  A  I  I  S  T  G  V  G  Q  H  Q  M  W  A  A  Q  Y  Y  K  Y  Y  R  K  P  R  Q  W  L

        1690        1710        1730        1750        1770        1790
ACATCTGGTGGATTAGGAGCAATGGGATTTGGTTGCCCGCTCTATTGGCGCGCTGTTGGAAGACCGGATGAAGTTGTGGTTGACATTGATGGTGATGGCAGTTTCATCATGAATGTG
 T  S  G  G  L  G  A  M  G  F  G  L  P  A  A  I  G  A  A  V  G  R  P  D  E  V  V  V  D  I  D  G  D  G  S  F  I  M  N  V

        1810        1830        1850        1870        1890        1910
CAGGAGCTTGCAACAATTAAGGTGGAGAATCTCCCAGTTAAGATTATGTTACTGAATAATCAACACTTGGAATGGTGTTCAATGGGAGGATCGGTTCTATAAGGCTAACAGAGCACAC
 Q  E  L  A  T  I  K  V  E  N  L  P  V  K  I  M  L  L  N  N  Q  H  L  G  M  V  V  Q  W  E  D  R  F  Y  K  A  N  R  A  H

        1930        1950        1970        1990        2010        2030
ACATACCTGGGGAATCCTTCTAATGAGGCGGAGATCTTTCCTAATATGCTGAAATTTCCAGAGGCTTGTGGCGTACCTGCTCCAAGAGTGACACATAGGGATGATCTTAGAGCTGCCATT
 T  Y  L  G  N  P  S  N  E  A  E  I  F  P  N  M  L  K  F  A  E  A  C  G  V  P  A  A  R  V  T  H  R  D  D  L  R  A  A  I

        2050        2070        2090        2110        2130        2150
CAGAAGATGTTAGACACTCCTGGCCATACTTGTTGGATGATTGACCTCATCAGGAACATGTTTTACCTATGATTCCAGTGGCCGGAGCTTCAAAGATGTGATCACAGAGGGTGAC
 Q  K  M  L  D  T  P  G  P  Y  L  L  D  V  I  V  P  H  Q  E  H  V  L  P  M  I  P  S  G  G  A  F  K  D  V  I  T  E  G  D

        2170        2190        2210        2230        2250        2270
GGGAGAAGTTCCTATTGAGTTTGAGAAGCTACAGAGCTAGTTCTAGGCCTTGTATTATCTCAAAATAAACTTCTATTAAGCCAAACATGTTCTGTCTATTAGTTTGTTGTAGTTTTTGCT
 G  R  S  S  Y  *

        2290        2310        2330        2350        2370        2390
GTGGCTTTGCTCGTTGTCACTGTTGTACTATTAAGTAGTTGATATTTATGTTTGCTTTAAGTTTTGCATCATCTCCCTTGGTTTTGAATGTGAAGGATTTCAGCAAAGTTTCATTCTCT

        2410        2430        2450        2470        2490        2510
GTTTGCAACATCCACTTGGTATCTGGAGATTAATTTCTAGTGGAGTTTTAGTGGCGATAAAATTAGCTTGTTGTTCACCATTTTTATTTCGTAAGCTATGTTGGGTCAGATTGGAAC
```

FIG. 4. Nucleotide and deduced amino acid sequences of the non-mutant allele of the tobacco *SuRA* gene (Mazur et al 1987).

1985; Fig. 3). This sequence homology prompted use of the yeast gene as a heterologous hybridization probe to isolate the ALS gene from tobacco. By this method, a DNA fragment encoding a protein of 667 amino acids was cloned from a genomic library prepared from tobacco plants homozygous for the *S4* mutation (Fig. 4; Mazur et al 1987). That this encoded protein is ALS was suggested by amino acid sequence homology with the *E. coli* and yeast ALS enzymes in scattered regions corresponding to the conserved domains of the microbial genes. RNA probes representing the 5' and 3' ends of this cloned tobacco gene hybridized to two distinct sets of fragments in fractionated enzymic digests of tobacco genomic DNA. One set of fragments corresponded to the cloned ALS gene. The other set of fragments confirmed the existence of a second ALS gene that had been identified by earlier genetic and biochemical studies (Chaleff & Bascomb 1987). The presence of two structural genes for ALS in the allotetraploid genome of *N. tabacum* was not unexpected. The question was, which of the two genes had been cloned from the resistant mutant: the *S4* allele of the *SuRB* gene or the non-mutant allele of the *SuRA* gene. Unfortunately, the herbicide-sensitivity of transgenic plants into which the cloned ALS gene had been introduced indicated that it was the non-mutant form of *SuRA* that had been recovered in this first effort (Mazur et al 1987). Subsequently, the normal allele of *SuRB* and mutant alleles of both loci were cloned and sequenced (K. Lee et al, unpublished results). These results demonstrated 0.7% divergence of the amino acid sequences encoded by the *SURA* and *SURB* genes.

It has been demonstrated previously that mutant alleles of yeast (*Saccharomyces cerevisiae*) and bacterial (*Escherichia coli*) ALS genes encoding herbicide-resistant forms of ALS contained single nucleotide differences that directed single amino acid substitutions in the enzyme (Yadav et al 1986). Presumably, in tobacco, resistance similarly results from mutationally directed changes of single amino acid residues in the ALS protein. If so, the cloning and transfer of the mutant alleles of the tobacco ALS genes can be expected to provide a powerful and efficient method by which to introduce herbicide resistance into presently susceptible crop species.

References

Chaleff RS 1983 Isolation of agronomically useful mutants from plant cell cultures. Science (Wash DC) 219:676–682

Chaleff RS, Bascomb NF 1987 Genetic and biochemical evidence for multiple forms of acetolactate synthase in *Nicotiana tabacum*. Mol Gen Genet 210:33–39

Chaleff RS, Mauvais CJ 1984 Acetolactate synthase is the site of action of two sulfonylurea herbicides in higher plants. Science (Wash DC) 224:1443–1445

Chaleff RS, Ray TB 1984 Herbicide-resistant mutants from tobacco cell cultures. Science (Wash DC) 223:1148–1151

Falco SC, Dumas KS 1985 Genetic analysis of mutants of *Saccharomyces cerevisiae* resistant to the herbicide sulfometuron methyl. Genetics 109:21–35

Falco SC, Dumas KS, Livak KJ 1985 Nucleotide sequence of the yeast *ILV2* gene which encodes acetolactate synthase. Nucl Acids Res 13:4011–4027

Mazur BJ, Chui CF, Smith JK 1987 Isolation and characterization of plant genes coding for acetolactate synthase, the target enzyme for two classes of herbicides. Plant Physiol (Bethesda) 85:1110–1117

Yadav N, McDevitt RE, Benard S, Falco SC 1986 Single amino acid substitutions in the enzyme acetolactate synthase confer resistance to the herbicide sulfometuron methyl. Proc Natl Acad Sci (USA) 83:4418–4422

DISCUSSION

Harms: You referred to the cost involved in screening chemicals in order to identify new herbicides versus the cost of developing crop genotypes that are resistant to an already existing herbicide. Isn't it true that in order to get a wide spread of performance in the field you have to convert a considerable number of crop genotypes to the desired herbicide resistant phenotype, whereas an innately selective chemical herbicide can be used for all genotypes? Therefore, the costs have to be multiplied by the number of crops that you have to convert in order to get the performance that you see with the chemical.

Chaleff: That's true, but the number of genotypes that would have to be made resistant would vary from one crop to another. Some crops, such as alfalfa, may have many varieties that are used in small local regions, but for other crops, such as corn, a smaller number of varieties tend to be used more widely. I think that if you were to introduce resistance into perhaps four or five widely used corn inbreds from which most of the hybrids are generated, you would be able to introduce resistance into a major portion of the crop with very few operations and comparatively inexpensively.

Galun: You mentioned a leader sequence on the gene for the yeast enzyme that was taken up into the chloroplast—was the leader a nuclear coded sequence for a chloroplast gene?

Chaleff: No, this is a mitochondrial leader sequence—the acetolactate synthase (ALS) yeast enzyme is localized in the mitochondrion.

Galun: And this leader sequence can be used to transport the enzyme into the chloroplast?

Chaleff: The yeast ALS gene was never transferred into plants so we don't in fact know if the yeast leader sequence can effectively direct the transport of the yeast enzyme into a plant chloroplast. But a leader sequence is found in the corresponding region of the plant ALS gene, i.e. at the 5′ end of the gene, that will presumably direct the encoded protein into the chloroplast. The yeast and the plant leader sequences are very different.

Zenk: Do these herbicide-resistant plants show any alterations in other general properties? For example, is yield the same in untransformed and transformed plants? Is plant growth the same, or do you observe differences?

Chaleff: Interestingly, in contrast to the case of atrazine resistance, we found no yield penalty associated with resistance in mutant plants, nor did we find any other associated morphological changes. In more detailed quantitative studies conducted by scientists of the R J Reynolds Company, no yield reduction was observed under field conditions.

Scowcroft: Because tobacco is a tetraploid, then with either one of the single mutants, *SurA* or *SurB*, there is still a normal acetolactate synthase gene. Is there a yield penalty in the double mutant?

Chaleff: No quantitative studies on dry weight and leaf weight have been done with that mutant. Measurements of shoot height suggest that there is no yield penalty. However, in the double mutant both mutations reside at or near the *SuRB* locus. So even in the *S4 Hra* double mutant (Fig.1), a non-mutant form of ALS (encoded by the *SuRA* gene) is present.

Potrykus: In your slide where you showed resistance in plants in the field, the wild-type plants are dying and the resistant plants are growing healthily, but at the back of the picture there were wild-type plants that looked healthier than the resistant ones.

Chaleff: Plants of both genotypes were of the same size. I can say so because those plants were harvested and the dry and fresh weights measured. It was those data that demonstrated that there is no yield penalty. Perhaps the angle of the photograph was misleading.

Hall: Are those differences in herbicide resistance for the genetically selected plants or for the gene transfer plants?

Chaleff: Genetically selected mutant plants that had undergone two back-crosses and then been made homozygous by self-fertilization.

Hall: Selection was on the herbicide (Glean)?

Chaleff: Transformation was checked against Glean, but transformants were initially selected on kananycin because the plasmid carried a gene for neomycin phosphotransferase.

Hall: Are these commercial varieties of tobacco?

Chaleff: No. I have only tested varieties of Xanthi, a Turkish variety which is not grown commercially in the USA. However, DuPont has an agreement with Northrup-King to develop the resistance commercially; plant breeders at Northrup-King are now backcrossing this trait of resistance into commercial varieties.

Hall: Did Glean go to any particular region of the plant or to any particular area within the plant cell?

Chaleff: In normal plants the herbicide is concentrated in the growing apical meristems.

van Montagu: You said that *SuRB* is not linked to *SuRA*. Then later on you proposed that *Hra* could be another mutation at the *SuRB* locus. If the *Hra* mutation were in the *SuRB* gene, how could you account for the result that *Hra* is linked to the independent mutation, *S4*?

Chaleff: The first mutation, the *S4* mutation, was in the *SuRB* locus. *Hra* was subsequently generated in that genetic background and was shown to be linked to *S4*. Therefore, it either resides within *SuRB* with *S4* or in another gene that is genetically linked to the *SuRB* locus.

van Montagu: You mean it was linked to *SuRB* and not to *SuRA*?

Chaleff: That is correct.

van Montagu: When you showed the effect of the herbicide on the plants, I got the impression that you didn't use concentrations which really kill the control plants. In our field tests with the Basta-resistant tobacco plants, we use conditions where the control plants are completely shrunken and dried out, while the engineered plants grow exactly like the wild-type plants. Did you use low concentrations of herbicide? We heard that Monsanto, in the case of their glyphosphate-resistant plants, probably cannot use the required concentrations of herbicide, otherwise they don't maintain full resistance and fertility of their engineered plants.

Chaleff: The impression that a sub-lethal concentration of the herbicide is being used is given by the characteristic mode of action of the herbicide. The herbicide will first inhibit plant growth without producing obvious symptoms of phytotoxicity or necrosis. Initially, the plant may even become greener and look very healthy, except for the apical region, which will appear chlorotic. The plant does not 'die' in the conventional sense for 2–4 weeks. That is a characteristic of this class of herbicide —plants don't look dead but they are. The herbicide concentrations that were used were actually very high and greatly in excess of those recommended for use in the field.

Harms: You referred to the atrazine-resistant mutant plants—this was a case similar to yours, in that the target enzyme or the target protein of the herbicide is changed from susceptible to resistant. Would you agree that these are the cases where a yield penalty is more to be expected than an example, such as Basta, where there is a detoxification mechanism that takes care of the herbicide before it ever gets in contact with the target?

Chaleff: Yes, I would think so; but it is always dangerous to generalize. I have presented an example of an altered target site that did not result in a yield reduction.

Scowcroft: What are the likely consequences of regular planting of a herbicide-tolerant tobacco cultivar on the development of resistance to that herbicide in weed species? What would the half-life of Glean or Ouse on a resistant tobacco plant? How long before weed species themselves mutate?

Chaleff: I can't project a figure for the lifespan of the herbicide or for the appearance of resistant weed species. I would expect resistant weeds to appear. If a mutation conferring resistance can arise in a crop one would expect similar mutations to arise in weeds in the field. To my knowledge, such resistant weeds have not yet appeared, but their absence probably reflects the short time that these herbicides have been in use. The continued application of a herbicide represents a very strong selective pressure and one can anticipate the appearance of resistant weeds. However, that does not necessarily mean that the herbicide is no longer of any use. By proper management and rotation procedures and by employing other classes of herbicides, one can eliminate the resistant weed population.

Zenk: The development of these plants involves considerable investment in time and money. How can you ensure in the long term that these plants are not used by competitors?

Chaleff: This is an interesting question and is a major concern for DuPont in commercializing these particular mutants. However, the spectrum of compounds to which these mutants display resistance does vary from one individual to the other. Certain resistant alleles do confer cross-resistance to a number of different compounds, even some with very different chemical structures. For example, some of the alleles confer cross-resistance against both sulphonyl ureas and imidazolinones, which are products of American Cyanamid and which have a chemical structure that is quite different from that of the sulphonyl ureas produced by DuPont and by CIBA-GEIGY. Other alleles confer a more selective form of resistance. For example, some mutations will confer a much higher degree of resistance to certain sulphonyl urea herbicides than to others. DuPont now is studying all of those effects and they are able to correlate particular mutational changes, that is, changes of specific nucleotides within the sequence of the gene, with resistance to particular compounds.

Riley: In the case of the transformed plants which you discussed at the very end of your paper, was the foreign genetic material inserted at one of the existing loci? If not, the dosage of non-mutant to mutant enzyme in the plant will be quite low. Can you comment on that?

Chaleff: The linkage studies to identify the site of insertion of the introduced gene haven't been done; I don't know how I could determine linkage to a gene encoding a sensitive form of the enzyme without having another linked marker. Unfortunately, we don't have a well-defined genetic map in tobacco and, therefore, mapping studies of the insertion have not been undertaken. Presumably, there are four copies of the non-mutant gene in a normal tobacco plant. Therefore, adding only a single copy of a gene encoding a resistant enzyme would result in only 20% resistant activity, if there is a strict gene-dosage relationship. In contrast, a plant homozygous for a resistance mutation at one of the two loci (i.e. two alleles of four being mutant) would possess 50% resistant activity. This is probably the reason why the transformants have a weaker phenotype than the mutants. Plants homozygous for a mutation at one of the two loci are completely resistant to 100 parts per million chlorsulphuron, whereas transformants display some inhibition of shoot growth at this concentration. One must also expect the degree of resistance expressed by transformants to be influenced by position effects, that is, that the expression of a foreign gene will depend on its site of insertion within the genome of the transgenic plant.

van Montagu: With *Agrobacterium*-mediated gene transfer there is no difficulty in selecting those plants that have two or three copies of the engineered genes. Those would then have the 50:50 ratio of mutated genes to normal ones. Approaches that would detoxify the herbicide, as we have done with Basta, will produce plants that have no yield penalty. So it is possible by using appropriate promoters and efficient genes to obtain perfect resistance.

Riley: I agree, but if what is being generated is parental breeding material, the more complex the genetic status of that material the more difficult it is going to be to transfer the new character into further derived lines.

Chaleff: There are other considerations: for example, the mutant allele that we are dealing with here confers such a high degree of resistance that even the observed partial resistance to 100 ppm of chlorsulphuron that is bestowed by a single introduced copy of the allele is more than ample to confer resistance in the field to conventionally used concentrations. Another point is that we can alter the gene promoters and thereby increase the rate of expression of the mutant allele without increasing the genetic complexity of the trait by introducing multiple copies of that allele.

Potrykus: From all that we know about the integration of foreign genes into the genome, it is highly improbable that it is integrated at the same locus. I am interested in the legal situation in the United States. If you introduce the foreign gene into a commercially available plant variety, do you create a new variety? Can you sell this new variety or do the rights belong to the owners of the original variety?

Chaleff: Presumably one can patent the new varieties generated by transformation and these varieties will be protected by patent law. Protection can be obtained under both the Plant Variety Protection Act and US patent law. There is currently much discussion about the relative advantages and disadvantages of each method of protecting genetically novel material. It might also be possible to patent the mutant gene itself and get commercial protection for that gene regardless of the plant into which it is introduced.

Potrykus: I was a little confused when you described this transformation. You said that you had problems in getting resistant plants by transforming these tobacco clones but you didn't give up until you had resistant plants. Do you have two clones, one clone from the wild type ALS and one from the mutant?

Chaleff: The first gene that was cloned we assume to be the wild-type allele of the *SuRA* gene present in the *S4* mutant background (the *S4* mutation resides at the *SuRB* locus). Subsequently, other libraries were screened, additional alleles of both the *SuRA* and *SuRB* genes were cloned and used in transformation experiments and resistance was obtained in that way. These cloned alleles have all been sequenced.

Hall: Have you determined the copy number of the native ALS gene in your Xanthi? and have you estimated the number of genes that were inserted in the mutant form?

Chaleff: These questions were answered by Southern blot hybridization analysis. The results of these studies indicated that two different genes are present in the wild-type genome. I don't know how many copies of the gene the Southern blots showed to be inserted in the transformants, but in all cases resistance did segregate as a single gene trait. So multiple insertions could have occurred at a single site, or additional copies of the gene could have been inserted at other sites but not expressed.

Ohyama: You didn't describe the mutation rate. You may develop a resistant crop but if the frequency of that mutation is high, then within a few years resistant weeds will arise and the cost of the development will be wasted.

Chaleff: I don't have a great deal of faith in efforts to measure frequencies or rates of mutation in cell culture because of certain technical considerations concerning the way in which cells grow in culture, such as their tendency to grow in aggregates. However, I have done seed mutagenesis studies that address your question. In soya bean, mutations conferring tolerance occurred at a rate of approximately 10^{-5} after mutagenesis. This number represents the mutation rate on a per genome basis calculated from the frequency of mutants in M2 seed populations (Sebastian & Chaleff 1987). However, none of these mutants had a resistant form of ALS and therefore resistance arose by some mechanism other than that reported here. Before extrapolating these results to natural populations, I urge you to consider that these mutations were induced and did not arise spontaneously and that they were recovered by deliberate screening of a soya bean population.

Cocking: So far we've discussed some of the problems that might or might not arise by adopting a transformation approach to herbicide resistance. Could you just say a little about the comparative merits of the different methods—there is the selection approach from tissue culture, the transformation approach and perhaps a mutation breeding approach. Which do you see as the best strategy for the development of herbicide resistance in crops other than tobacco?

Chaleff: I think that, as you have indicated, the choice of strategy would depend on the individual case. *In vitro* selection and transformation both require passage through cell culture; in species for which plant regeneration from cell culture is difficult or has not yet been accomplished, I would therefore opt for mutation breeding, soya bean being a good example. In choosing between transformation and *in vitro* selection, again in some cases you can regenerate plants from cell culture but not from cells that are capable of being genetically transformed. For example, in maize successful transformation has only been reported in one case and even then fertile plants were not recovered. So although transformation techniques are just being developed for maize, mutants resistant to the imidazolone herbicides were successfully selected in cell culture and mutant plants regenerated. These maize mutants are now being commercially developed in a joint venture between American Cyanamid and Pioneer Hi-Bred Seed Company. Where one can accomplish transformation and *in vitro* mutant selection with equal facility the question is more difficult. I would opt for transformation, despite the larger initial investment of time in cloning the mutant gene. Again this choice presumes that one can clone the gene, that is another technical obstacle that would have to be overcome. An important advantage of transformation is that it utilizes a genetically characterized gene which confers resistance of a predictable type. One doesn't have to rely upon random mutation to generate the desirable phenotype independently for each separate crop. This feature of transformation addresses the point raised by Professor Zenk with respect to cross-tolerance to commercially competitive herbicides. Ideally, one wants to introduce a mutant allele that confers resistance selective to your chemistry: that is the allele that you want to clone and introduce into all crop varieties. Recovery of that same specificity independently in each crop by means of random mutation would be a most tedious process. Returning to Dr Harms' point about the number of commercial varieties of individual crops that exist, I think that such considerations identify another important advantage of transformation. It is far easier and faster to introduce resistance into a large number of varieties by transformation than by repetitive selection or by very expensive cross-breeding and inbreeding programmes.

Zenk: Has acetolactate synthase, either from microorganisms or from a plant, been crystallized, or does any of the molecular work you have done give any information about the site of inhibition? Could you predict a rational design for future synthesis of herbicides?

Chaleff: That's a fascinating area of research that I would like to see pursued, but to my knowledge isn't being investigated effectively anywhere. One of the advantages of cloning a plant ALS would be to provide an ample source of the enzyme for crystallization and subsequent structure-activity relationship studies. This strategy could lead a chemical company into a very useful programme for rational development of herbicides.

Potrykus: Your strategy should produce amplification. Now that you have a probe for your gene, are you looking for amplification?

Chaleff: We never obtained amplification, although we have looked for it. The criterion for amplification in our screens was increased levels of the enzyme; that criterion might be inappropriate for detecting amplification, if the amount of enzyme is controlled post-transcriptionally. We haven't yet examined all of our mutants by Southern blots to detect amplification.

Yamada: You are fortunate to work on an enzyme such as ALS, which is involved in valine, leucine and isoleucine biosynthesis. I don't know which pathway has been inhibited by the herbicide in your case, but I think you are very wise to choose this enzyme. Did you intentionally choose this system or was it fortuitous?

Chaleff: After such flattering comments about my wisdom it is difficult to admit that my choice was fortuitous, but it was—DuPont was working on these herbicides when I went there. My motivation at the time was not to use cell culture to produce herbicide-resistant plants, but to use the techniques of biochemical genetics applied through cell culture to identify the mode of action of these herbicides. That approach did justify itself. It was by selecting and analysing mutants that we were able to identify the altered site and therefore the site of action of the herbicides which previously was unknown. I had sceptically considered that such mutations would have deleterious effects on either the catalytic efficiency of the enzyme or the vigour and yield of the plant.

One can rarely predict the full consequences of mutation and I think that for this reason we must always go ahead and do the experiment with the utmost optimism without allowing all the projections of why an experiment can go wrong to prevent us from doing that experiment.

Barz: You described a system where the resistance is provided by an ALS mutation. There are other cases where the resistance is brought about by metabolism of the herbicide. In the case of the sulphonyl ureas, such metabolism involves hydroxylation and subsequent glucosylation. How would you compare the efficiency of resistance through differential or preferential metabolism to that of your system?

Chaleff: The inherent tolerance for sulphonyl urea herbicides of some crop plants does seem to be primarily, if not exclusively, dependent on metabolism of the herbicide, but the degree of such tolerance can't compare with the resistance that we obtain by mutations in the gene encoding the ALS enzyme. The levels of herbicide to which these mutants are resistant would destroy crops that are tolerant on the basis of their ability to metabolize the herbicide.

Barz: But there is a good chance that weeds will also mutate, whereas metabolism requires additional enzymes and it is less likely that weeds will develop such new enzymes. The specificity or the permanence of this resistance will be higher, if it is based on metabolism.

Chaleff: Yes, but even in fields where the tolerance of the crop is due to metabolism of the herbicide, the ALS gene within the weeds can still mutate. The weeds are not concerned with the mechanism by which tolerance is achieved in the crop.

Yamada: I would like to introduce one of our results from our work on photoautotrophic cells. Unfortunately, this was not a cereal plant, but tobacco as usual. We selected a cell line resistant to a photosynthesis-inhibiting herbicide, atrazine. As Dr Chaleff has described, recent advances in plant cell culture have facilitated the introduction of herbicide-resistance into crop plants. But success has been limited to those herbicides whose modes of action result from interference with non-photosynthetic processes. In the United States and Europe only some weeds have been selected for their resistance to photosynthesis-inhibiting herbicides. By culturing tobacco cells photoauto-trophically, we obtained one cell line that is completely resistant to 30 µM atrazine. The growth of our resistant cell line was even enhanced slightly at a lower atrazine concentration (1 µM). We were very curious about this cell line, so we sequenced the DNA of the Qβ protein, the target protein of this herbicide. Drs Hirshberg and McIntosh (1983) reported that in the intact resistant amaranthus plant, the serine at position 164 has been changed to glycine by mutation. In our selected cell line, this serine has mutated to threonine (Sato, Shigematsu and Yamada, unpublished work). I think that such selection could be a very useful strategy to obtain photosynthetic inhibitor-resistant plants from photoautotrophic culture.

Galun: How is the photosynthetic efficiency in the absence of atrazine, is it also less efficient than in the wild-type?

Yamada: Our photoautotrophic cultured cell line, which comes from C3 plants, has higher PEP carboxylase activity than Rubisco. This is most unusual. We were very happy because we thought that we would be able to regenerate a plant with much higher photosynthetic ability, but so far we have not been able to regenerate any plants. The photosynthetic activity of our photoautotrophic cells is slightly lower than that of the mesophyll cells of tobacco.

Galun: Could you transfer the chloroplast by protoplast fusion into tobacco?

Yamada: We first tried to culture the chloroplasts themselves, but we never succeeded. We are now trying to transfer the chloroplasts to non-chlorophyllous cells.

Fowler: The increase in PEP carboxylase activity, is it a single enzyme, an isozyme or what?

Yamada: I don't know.

Barz: It is not an isozyme. The high PEP carboxylase activity is observed only at a phase in the cell culture when there is a high rate of cell division. As the culture approaches the stationary phase, the PEP carboxylase activity declines and Rubisco enzyme activity increases. Therefore, the PEP carboxylase in this system has the function of an anapleurotic enzyme.

Fowler: Does it show any signs of activation by bicarbonate or pyruvate?

Barz: Not by bicarbonate, but you can influence the ratio of PEP carboxylase to Rubisco by the ratio of ammonia to nitrate—the higher the ammonia concentration in the medium, the higher the PEP carboxylase activity.

Fowler: So presumably you have a C4 acid trapping system into amino acid biosynthesis.

Barz: Exactly, that's the function of an anapleurotic enzyme at times of rapid cell division. This refers to the fact that chlorophyll synthesis and the rate of chloroplast multiplication are not in synchrony with cell division.

Potrykus: Dr Chaleff, what are your feelings on the public concern about herbicide resistant plants. In discussions with students, it is something about which they are very worried—that big companies are producing herbicide-resistant plants and are spraying more herbicides onto fields.

Chaleff: My own view as a private citizen rather than as a company employee is that herbicide resistance introduces more possibilities to make environmentally safer herbicides. In the past, one of the considerations in developing herbicides was that one had to develop compounds that had inherent within their chemistry the selectivity that would spare the crop but kill the weeds. That's a very difficult property to design; it is a formidable task that I myself would not want to undertake and I'm amazed at the success of the chemical industry in achieving such selectivity. If the criterion of selective activity is removed from the chemistry, one can simply produce potent but environmentally safe compounds without worrying about their specificity of action and then hand this compound to the geneticists and ask them to introduce the desired selectivity by genetic means. I think we are on the threshold of a new era for cooperativity between the chemists and the geneticists in developing much safer chemicals for use as herbicides. I think that modern agriculture depends on herbicides and it would be ludicrous to propose that we abandon the use of these chemicals in modern agriculture. It seems more reasonable to propose that the powerful methods of biotechnology be employed to assist chemists in designing herbicides that are safer both for the environment and for animal life.

Potrykus: Total herbicides which kill every plant unselectively are of even greater concern.

Chaleff: My concern is for the use of chemicals in very large concentrations, such that these compounds cannot be removed quickly enough from the environment but penetrate ground water supplies. In the sulphonyl ureas we have examples of compounds that can be used at one-hundredth of the conventional application rates. Levels of grams per hectare are remarkably low and minimize contamination of the environment. Future generations of herbicides produced in concert with genetic modification of the crop may be even more effective.

Riley: Of the revolutions that have occurred in agriculture in the last hundred or so years, first was the use of fertilizers, second the use of selective herbicides and third the use of plant breeding. Herbicides are crucial to modern agriculture and improvements in their efficiency in these ways are of great significance.

References

Hirshberg J, McIntosh L 1983 Molecular basis of atrazine resistance in *Amaranthus hybridus*. Science (Wash DC) 222:1346–1349

Sebastian SA, Chaleff RS 1987 Soybean mutants with increased tolerance for sulfonylurea herbicides. Crop Sci 27:948–952

Somaclonal variation

William R. Scowcroft* and Philip J. Larkin†

Biotechnica Canada Inc., Calgary, Canada* and Division of Plant Industry, CSIRO, Canberra, Australia†

Abstract. Eukaryote genomes are in a dynamic state of flux. This is most apparent in higher plants under *in vitro* culture where the amount of variability generated is extensive. Genetic and molecular analysis is providing some understanding of the events which give rise to somaclonal variation. In maize, a new fully functional electrophoretic variant at the alcohol dehydrogenase locus has been found among somaclones and subsequently characterized as resulting from a single nucleotide substitution. In wheat, variants have been analysed which affect traits such as height and alcohol dehydrogenase synthesis, as well as complex gene loci involved in the synthesis of grain amylases. Somaclonal analysis in maize and alfalfa has shown that cell culture greatly enhances the activation of transposable elements. For plant improvement, the enhanced frequency of genomic rearrangements during culture provides a new option to introgress alien genes from wild relatives into domesticated crops, and may enable the products of 'new' types of mutational events to be recovered.

1988 Applications of plant cell and tissue culture. Wiley, Chichester (Ciba Foundation Symposium 137) p 21–35

The original concept of somaclonal variation (Larkin & Scowcroft 1981) has gained widespread acceptance and has been the subject of numerous reviews (see Scowcroft 1985, Larkin 1987). This variation, which arises as a consequence of tissue culture, has been found in essentially all plant species that have been regenerated from tissue culture. More recently, research has been directed at understanding the cause of somaclonal variation rather than mere documentation of the phenomenon. Several critical analyses have been conducted at the genetic, cytogenetic and molecular levels. These analyses reveal that somaclonal variation can give rise to mutants which behave as classical Mendelian mutants, genetic changes which affect characters under the control of multigene families, those which affect polygenically determined traits and those which give the appearance of resulting from activation of transposable elements. Polyploid and aneuploid variants occur frequently, as do chromosomal rearrangements such as duplications, deficiencies, translocations and other interchanges.

The following discussion provides some details of several recently analysed somaclonal variants which give some insight into the nature of the genetic changes that occur during tissue culture.

Mendelian mutants

The genetic analysis of plants from tissue cultures of diverse species such as rice, wheat, maize, lettuce, tobacco, tomato, celery and rapeseed, demonstrates the occurrence of classical point mutations. Many of these mutants are allelic to other known mutants. In many cases, there is substantial evidence that these mutations occurred during the tissue culture phase because sister plants regenerated from the same callus were non-mutant. One such mutant, which has been analysed at the nucleotide level, arose in regenerated maize plants which were screened for isozyme variants at the alcohol dehydrogenase (*Adh1*) locus (Brettell et al 1986a).

Maize alcohol dehydrogenase mutant

Plants regenerated from an F_1 between the allelic variants, *Adh1-S* and *Adh1-F* were screened for isozyme variation by electrophoresis. Among 645 primary regenerants, several showed presumptive variation but only one (*Adh1-Usv*) was stably transmitted in subsequent seed generations. This mutant allele produced a functional enzyme with a slower electrophoretic mobility than either one of its progenitor alleles. This mutant allele was cloned using an ADH genomic DNA probe derived from the *Adh1-S* allele. Subsequent restriction endonuclease analysis showed no major sequence reorganization and also confirmed that the mutant allele was derived from the parental Adh1-S allele rather than the fast-migrating allele.

Sequencing of the entire mutant *Adh1-Usv* gene revealed a single base change in the coding triplet of exon 6 immediately adjacent to the junction with intron 6. This substitution of thymine for adenine translates into a polypeptide sequence where a glutamic acid residue is replaced by valine. This, in turn, significantly affects the charge on the ADH molecule resulting in slower electrophoretic mobility.

The spontaneous rate of mutation at the *Adh* locus is very low, i.e. 10^{-5} – 10^{-6}. Although only a single mutant has been found as a result of somaclonal variation, the frequency at which it occurred was in the neighbourhood of 10^{-3}.

Somaclonal mutants and chromosomal rearrangement

One of the most frequent causes of somaclonal variation is chromosomal rearrangement. Larkin (1987) documents numerous examples of recorded evidence for deletions, chromosome fusion and interchanges. The consequences of such rearrangements can result in mutations which affect the phenotypic expression of one or more genes.

In wheat, Larkin et al (1984) reported extensive genetic variability among somaclonal regenerants and simultaneously Karp and Maddock (1984) reported frequent occurrence of chromosome variation in wheat plants regenerated from tissue culture. In a recent analysis, Davies et al (1986) analysed the progeny of more than five hundred plants regenerated from the wheat cultivar, Millewa, for potential variation in the expression of alcohol dehydrogenase-1 isozymes. *Adh1* in hexaploid wheat is complicated by the fact that the locus is triplicated and that functional ADH isozymes are dimers formed by random association of each of the three different monomers.

Seventeen ADH variants were identified and analysed. Simple aneuploidy for chromosomes 4A or 4D accounted for 13 of these variants and the remaining four were euploid. A genetic and cytogenetic analysis of three of the mutant euploid somaclonal variants indicated that three different types of chromosomal rearrangements had occurred. The translocation mutant, *SV1*, resulted from a translocation of part or all of the chromatin from 4Aα to the short arm of chromosome 7B. The translocated segment carried the *Adh-A1*, *Gai1* (gibberellic acid insensitive) and *Rht1* (reduced height) loci. This stable, non-reciprocal translocation line now has four doses of the *Adh-A1* and *Rht1/Gai1* loci rather than the normal two. The phenotypic consequences of this are an altered alcohol dehydrogenase isozyme pattern and plants which are substantially reduced in height.

A second mutant, *SV5*, was also shown to be the result of a non-reciprocal translocation. A segment of the short arm of chromosome 3B had been translocated to chromosome 4Aα resulting in the loss of function of the *Adh-A1, Rht1* and *Mslc* (male sterility) loci.

The third mutant, *Iso 4Aα*, presumably resulted from breakage and centric fusion, to produce an iso-chromosome. The phenotype reflected increased dosage of both *Adh-A1* and *Rht1* genes.

The reason why cell and tissue culture result in an increased frequency of chromosomal aberration is yet to be explained. Several authors, but most notably Benzion et al (1986), argue that late-replicating heterochromatin 'may occasionally replicate so late in mitosis of cultured cells that bridge formation and subsequent chromosome breakage occurs at anaphase'. This would lead to a breakage-fusion-bridge cycle with consequent genetic effects, such as deletions and chromosome interchanges.

Altered gene copy number

Plant genomes can be large, with as much as 60% of the genome composed of repeated DNA. Repeated DNA can have a reiteration frequency per sequence as high as $10^5 - 10^6$ and can be clustered as tandem arrays or dispersed throughout the genome.

The ribosomal RNA (rRNA) genes of plants represent such highly repeated DNA sequences. Brettell et al (1986b) examined the consequences of somaclonal variation in regenerated triticale plants by screening for possible perturbation of rRNA genes.

From among 192 regenerated plants of triticale, screened by Southern hybridization with a rDNA spacer probe, one variant displayed substantial perturbation. This variant showed an 80% reduction in the rDNA spacer sequences of rye chromosome 1. This sequence depletion, detected initially by the rDNA spacer probe, also had a substantially reduced level of C-band staining as revealed by cytological analysis. This rDNA-depleted locus is the site of the nucleolar organizer in the rye genome component of triticale.

'New' genes from somaclonal variation

The question is frequently put: 'Among the array of somaclonal variants derived from tissue culture, are the mutants the result of changes at pre-existing loci or do entirely new mutants occur?' Certainly the former is true but the evidence for entirely new mutants is, at best, only circumstantial.

Some genetic events, which occur at a very low rate spontaneously, occur far more frequently during tissue culture and for this reason give the impression that somaclonal variation 'creates' new genes.

Evidence is already accumulating that transposable element activity is greatly enhanced during tissue culture. In maize, activity of the Activator component of the classical Ac-Ds transposable element system is greatly enhanced in plants regenerated from maize tissue culture (Benzion et al 1986). In alfalfa, Groose and Bingham (1986) reported the occurrence of an unstable anthocyanin mutation which occurred in plants regenerated from tissue culture.

If enhanced transposable element activity is a consequence of tissue culture, then several opportunities exist for seemingly 'new' genes to appear. Transposition of DNA sequences can relieve genes from pre-existing repression of activity, or excision of an element from a structural gene can restore function to the gene product. Transpositions can also move structural gene sequences into regions where they fall under the control of different promoters, which could conceivably alter the level and developmental timing of gene expression.

A somaclonal variant has recently been described which affects the isozyme pattern of one of the seed protein complexes of wheat, namely β-amylase (Ryan & Scowcroft 1987). The mutant phenotype appears to be entirely new. This variant was found among 149 regenerants and was characterized by at least five new isozyme bands, as well as an increased intensity of two previously existing bands. This variant segregated with Mendelian fidelity, no recombination has been observed between the variant bands, and mitosis and meiosis are cytologically normal both in the homozygous variant and in the F_1

between the variant and its parent. The variant pattern has not been observed in a survey of more than 100 diverse wheat genotypes which represent the 10 known β-amylase phenotypic groups. This new variant has not previously been observed and therefore gives the appearance of being a new gene. It could represent a rare mutation leading to the expression of a previously silenced locus.

Application of somaclonal variation

Since its inception, somaclonal variation has been regarded as a new source of genetic variability for plant improvement (Larkin & Scowcroft 1981, Evans & Sharp 1986, Bingham & McCoy 1986, Ryan et al 1987). One of the major potential benefits of somaclonal variation is the creation of additional genetic variability in co-adapted, agronomically useful cultivars, without the need to resort to hybridization. This takes on added appeal if *in vitro* selection is possible or if rapid plant screening procedures are available. Characteristics for which somaclonal mutants can be enriched during *in vitro* culture include resistance to disease pathotoxins and herbicides, and tolerance of unfavourable environmental or chemical conditions.

Somaclonal variation for other characters for which there is no adequately defined *in vitro* response, such as seed protein quality, photosynthetic efficiency or yield, can be of benefit, provided effective whole plant screening protocols are available.

A new option to introgress alien genes from wild relatives into crop species has emerged from the observation that chromosome rearrangements occur in plants regenerated from tissue culture. The enhanced level of chromosome exchange in tissue culture might be employed to effect alien gene transfer. In crops such as wheat, wide crossing programmes have produced alien chromosome addition lines following hybridization with relatives such as *Aegilops*, *Agropyron*, *Elymus*, *Haynaldia* and *Hordeum*. These alien species carry genes for resistance to wheat disease pathogens and tolerance to environmental stresses such as heat, salinity and metal toxicity. Research has been initiated to utilize tissue culture to enhance the rate of transfer of specific genes from these wheat alien addition lines. This could emerge as one of the most salutary consequences of somaclonal variation for plant improvement.

In the horticultural and forestry industries, micropropagation is being used increasingly as a production process to rapidly propagate elite genotypes. Somaclonal variation is a serious disadvantage to such operations because clonal uniformity may not be sustained. It is very important therefore that the mechanisms which give rise to somaclonal variation are understood, so that procedures can be developed to mitigate against this unwanted source of genetic and phenotypic variability.

References

Benzion G, Phillips RL, Rines HW 1986 Case histories of genetic variability in vitro: oats and maize. In: Vasil IK (ed) Cell culture and somatic cell genetics of plants. Academic Press, New York, vol 3

Bingham ET, McCoy TJ 1986 Somaclonal variation in alfalfa. Plant Breed Rev 4:123–152

Brettell RIS, Dennis, ES, Scowcroft WR, Peacock WJ 1986a Molecular analysis of a somaclonal mutant of maize alcohol dehydrogenase. Mol Gen Genet 202:235–239

Brettell RIS, Pallotta MA, Gustafson JF, Appels R 1986b Variation at the *Nor* loci in triticale derived from tissue culture. Theor Appl Genet 71:637–643

Davies PA, Pallotta MA, Ryan SA, Scowcroft WR, Larkin PJ 1986 Somaclonal variation in wheat: genetic and cytogenetic characterization of alcohol dehydrogenase 1 mutants. Theor Appl Genet 72:644–653

Evans DA, Sharp WR 1986 Applications of somaclonal variation. Bio/Technol 4:528–532

Groose RW, Bingham ET 1986 An unstable anthocyanin mutation recovered from tissue culture of alfalfa (*Medicago sativa*). Plant Cell Rep 5:104–107

Karp A, Maddock SE 1984 Chromosome variation in wheat plants regenerated from cultured immature embryos. Theor Appl Genet 67:249–255

Larkin PJ 1987 Somaclonal variation: history method and meaning. Iowa State J Res 61:393–434

Larkin PJ, Scowcroft WR 1981 Somaclonal variation — a novel source of variability from cell culture for plant improvement. Theor Appl Genet 60:197–214

Larkin PJ, Ryan SA, Brettell RIS, Scowcroft WR 1984 Heritable somaclonal variation in wheat. Theor Appl Genet 67:443–455

Ryan SA, Scowcroft WR 1987 A somaclonal variant of wheat with additional β-amylase isozymes. Theor Appl Genet 73:459–464

Ryan SA, Larkin PJ, Ellison FW 1987 Somaclonal variation in some agronomic and quality parameters in wheat. Theor Appl Genet 74:77–82

Scowcroft WR 1985 Somaclonal variation: the myth of clonal uniformity. In: Hohn B, Dennis ES (eds) Genetic flux in plants. Springer-Verlag, Berlin, p 217

DISCUSSION

Potrykus: I was surprised that you don't see much application for somaclonal variation, because I always understood that you saw lots of applications.

Scowcroft: You have to understand what I mean by its application. The simple procedure of cell culture followed by extensive plant regeneration and then screening for useful variability among the progeny of those plants is probably a forlorn hope. The use of somaclonal variation in this sense is akin to mutagenesis. There are a couple of exceptions where the breeders have no real source of genetic variability, such as banana and some root crops, because these are obligate asexually propagated species. I think you need either a very good screen at the plant level or a very good selection system in cell culture to use somaclonal variation directly.

Potrykus: Variation was observed long before it was termed 'somaclonal variation'. Is it settled now when this variation occurs? Is this variation which you have described a consequence of tissue culture or does it occur in normal plant development?

Scowcroft: In most cases, given that the control experiments are done properly, you can distinguish between whether the somaclonal variation is pre-existing or whether it has occurred during tissue culture.

Potrykus: How do you do the control experiments properly?

Scowcroft: In all of those cases I described, none of the material we looked at involved less than five regenerated plants from a single explant—in other words there was a set of sister somaclones. We did find some cases where there were sister somaclones which were also the same as the initial mutant we had found—we discarded those because they may have been pre-existing.

Potrykus: Did you find sister somaclones which indicated that this variation existed before you put this explant into culture?

Scowcroft: Yes, we did.

Potrykus: If you are sure that this variation existed before you started culture, then the question is, how much of the variation you described is due to this pre-existing variation?

Scowcroft: Again this depends on doing the right controls. One is this production of sister somaclones for comparison. The other one is that selfed seed from the original plant from which the explant was taken is retained for subsequent comparison. In the case of wheat, we took immature embryos from a maturing head; additional heads on that plant were selfed and the parental seed saved. When a variant was detected, we went back to the original parent and asked whether the variation was present in the selfed seed derived from the parent. This is important because monosomics occur frequently in wheat.

Galun: We recently looked at the origin of somaclonal variation using DNA probes from the *Nor* spacer region on Southern blots to determine whether the variation was pre-existing or really occurred as a result of scutella callus formation and shoot regeneration. The answer is quite complicated. There are some varieties where there is barely any detectable somaclonal variation. In some specific varieties, the varation is exceedingly high—so it's a sporadic occurrence (Breiman et al 1987). Recently we retested somaclonal variation in this systen by taking single spikes and dividing each spike into two parts: the upper part was used for embryo culture and the lower part was retained for seeds. Each spike was tested individually to determine how much variation there is between spikes of the same plant and how much variation occurred in the future callus-derived plants and thus is probably caused by somaclonal variation. In some plants there was no variation between spikes. Furthermore, among plants regenerated from the scutellar calli derived from certain spikes there was tremendous variation—Southern blots showed the appearance of new *Taq*I restriction digest patterns (T. Felsenburg, A. Breiman, E. Galun, unpublished work). Therefore, to refer to a percentage variation is actually saying very little because this is caused by tremendously high variability in individual cases. It is possible that there is a gene which promotes somaclonal variation.

Scowcroft: There may be genes or genotypes which promote the occurrence of somaclonal variation. Ron Phillips proposed a set of experiments to address this, but the results are not available yet. He has the notion that somaclonal variation is a function of late-replicating DNA. This makes sense theoretically: if DNA replication and cell division are out of synchrony, bridges are formed during mitosis which result in a breakage-fusion-bridge cycle. This can lead to exchanges, interchanges or translocations. If you read Barbara McClintock's work, this is the reason why transposable elements exist. Phillips' experiment is utilizing chromosome 10 of maize, where they have stocks of different amounts of heterochromatin which is late replicating. I think their belief is that most of the variation that is coming out of tissue culture is a consequence of late-replicating DNA, i.e. the DNA has not completed replication before cell division occurs.

Galun: In our experiments, the plants had a completely normal cytology (T. Felsenburg 1987 PhD thesis, Weizmann Institute of Science).

Fowler: If we move from crop plants to cell culture, those of us who work in the latter area and with natural products face a different type of dilemma. On the one hand somaclonal variation could be very useful in enabling us to select and develop high-yielding cell lines. On the other hand it poses major problems for the maintenance of high-performing lines, once they have been selected. I have always held the view that we should perhaps not be surprised at the presence or indeed levels of instability in cell cultures. After all, these represent a highly aberrant and stressful situation for the individual cells of the culture, away from the highly integrated and balanced environment of the whole plant. The question to our geneticist colleagues is do they have any ideas about the mechanisms underlying this instability and whether or not it might be controllable.

Scowcroft: I don't think I can offer you much comfort. In any sort of bioreactive fermentation, you will have to test that your material continues to do what it was selected to do; you may have to regenerate those cultures from time to time.

Fowler: In essence, we are in the same position as those working with classical microbial antibiotic fermentations, where there is a constant selection system.

van Montagu: With the first mutant that you described, *Adh1-Usv*, where glutamic acid had been changed to valine, did you do Northern blots to see if this transversion affected the processing of the transcript?

Scowcroft: No, we didn't, but we presume it's being processed quite normally.

van Montagu: Could it be that there is another mutation that results in higher enzyme levels?

Potrykus: We see a general flexibility of the plant genome which is quite interesting and worth studying, even if there is not much hope of exploiting it commercially. I think basic studies are lacking and it is important for those who are studying somaclonal variation to get really clear data about the effect— when it is occurring, how it is occurring, whether it contributes to regulation of developmental phenomena. Somaclonal variation has aroused interest in lots of groups and companies, and many people are investing lots of money in trying to use it for commercial purposes. If you had to advise somebody whether or not to invest money in trying to use somaclonal variation for any commercial purpose, what would you advise?

Scowcroft: A number of companies are putting effort into using somaclonal variation to develop high solids tomato—I think they are wasting their money because the genes are already available. The company I work with is trying to use somaclonal variation to facilitate transfer of disease resistance into breeding material from alien species, because there is no other way to achieve this efficiently. Intra-specific crosses have been made, backcrosses have been subsequently done, which result in an alien chromosome substitution that confers immunity to black leg tolerance but is unstable. Somaclonal variation could result in introgression of part of the chromatin carrying the gene into the host genome.

Potrykus: Would a more feasible alternative be to break up the chromatin by heavy irradiation and fuse these protoplasts with your *Brassica* protoplasts?

Scowcroft: I don't think that is a feasible alternative. Such a shotgun approach to chromosome rearrangement is unlikely to get you anywhere. Why are you doing protoplast fusion anyway? You already have that chromosome or part of the chromosome in its right background. The objective is to introgress part of it into the host genome by rearrangement. The other aspect of somaclonal variation is that the results can have serious economic consequences. A case in point is the experience of Unilever with oil palm, where plantations were established with plants derived from tissue culture. A significant number of the plants were sufficiently sterile to render the plantation uneconomic. The sterility problem resulted from chromosomal abberations.

Galun: They used a long period of callus growth, after which a high percentage of palms were sterile. One doesn't know what would have happened, had they used a shorter period of callus growth.

Fowler: I understand that another problem relates to hormonal gradients that have been applied to those systems. It may not be simply the long period as a callus, but partly an effect of 2,4-D, naphthylacetic acid and other plant growth regulators, and the way in which those have been applied. This reflects the complexity of integration of control systems which go beyond the single cell. My concern is the lack of understanding of the multicellular integration and the control systems involved.

Riley: Bill, are you saying that the variation which is generated somaclonally is no different from the variation that can be generated by other means, such as X-irradiation? Because you didn't describe anything which was different from what can be produced conventionally.

Scowcroft: It's a question of what is meant by different. I think that the spectrum of mutations obtained from somaclonal variation is probably no different to those derived from a chemical mutagenesis or a high energy radiation mutagenesis programme. But I believe that the causes of those mutations are different. Let's say, for example, that it is late-replicating DNA that is generating this sort of variation. That is not the usual cause of mutations using conventional mutagenesis techniques. So the mechanism which generates somaclonal variants is different to that in other sorts of methods but you still get the same spectrum of mutation.

Riley: There are examples of late-replicating DNA in normal development in normal plants that gives rise to variation, so late replication is nothing unique.

Scowcroft: I agree, but it is occurring at a higher frequency.

Chaleff: On the basis of the data that you presented, how important can late chromosome replication and a breakage-fusion-bridge cycle be to the changes you observe occurring spontaneously in cell cultures? You said that you regenerated 540 plants randomly and screened them for electrophoretic variation in ADH and detected such variation in only one plant. That variation in electrophoretic mobility represented a single nucleotide change and therefore a point mutation, which is not usually thought to arise by breakage-fusion-bridge formation. Moreover, the only types of point mutations you could have detected in such a system would be those that altered the charge of the protein. Presumably many more electrophoretically neutral changes were occurring that went undetected. Therefore your point mutation frequency had to be very high to be detected by screening for variation in electrophoretic mobility. It would seem to me that there is another mechanism of mutation operating in cell culture that is far more important than the breakage-fusion-bridge cycle in generating the events that you observed.

Scowcroft: The electrophoretic variant you refer to was recovered from maize regenerants. Maize is a diploid and probably cannot tolerate gross chromosomal aberrations.   Surprisingly, we didn't find any null allelles which could have resulted from small deletions. On the other hand, in wheat, which is a hexaploid species, the 16 mutations affecting ADH all resulted from chromosomal abnormalities.

Riley: Bill, you spoke about the 4A:7B translocation in wheat and said that it wasn't reciprocal, but you showed a ring of four chromosomes at first metaphase in a heterozygote, therefore it must be reciprocal—the only way you can get a ring is by having a reciprocal translocation.

Scowcroft: Certainly that is true.

Riley: What that means for your segregation ratio, I don't know; it needs to be worked out.

Tabata: Dr Scowcroft proposed a rational explanation for somaclonal variation, showing some examples of somaclonal variations caused by rather simple cytogenetic changes. But in cultured cells we often observe a wide continuous variation in quantitative characteristics, such as the amounts of secondary products formed, in which the productivity can be greatly improved by selection of cellular clones. Would you expect that this kind of variation might be caused by amplification of DNA? Do you have any examples of such cases?

Scowcroft: I don't have any examples from our own work but there is some circumstantial evidence. In a number of cases, which include *in vitro* selection for herbicide tolerance (phosphinotricine and glyphosphate), amplification of particular structural genes has resulted. It can be argued that the amplification of particular genes was a consequence of somaclonal variation.

Potrykus: Amplification, as detected using DNA probes, is a quite frequent event in culture. We have looked at amplification in a specific case. We produced a number of clones that had different levels of amplification because we wanted to study the fate of amplified DNA during plant development and especially during meiosis. In none of the amplified clones could we get plant regeneration, although we were working with a very regenerable system. Our general impression is that amplification of DNA in some way interferes with plant development (M.W. Saul, personal communication). This may be an explanation for the cases where people have isolated an increase in productivity of a certain product, then lost that characteristic during plant regeneration.

Zenk: An observation was made by Dr E. Kolossa in our laboratory in a PhD thesis using a strain of *Nicotiana tabacum*. Cell suspension cultures were prepared, then plated out and the content of nicotine measured by radioimmunoassay. We selected clones which contained up to 1% nicotine and some which had no nicotine detectable by the radioimmunoassay. We screened several thousands of these clones and selected about 20 high-producing and 20 low-producing clones and regenerated them. To our surprise, we saw that these chemical characteristics were completely lost in the differentiated plants.

Tabata: We once regenerated plants from a high nicotine-producing cell line and this plant had normal nicotine content, but when we induced callus tissue again from this plant the callus tissue had a very high nicotine content (Tabata 1978). I think this character is transmitted through the cytoplasm and does not express itself in the whole plant but only in the cultured cells. There must be something strange that we cannot explain going on in cultured cells. We should consider also the unequal distribution of cytoplasmic factors, for example, cell organelles participating in secondary metabolism.

Galun: People working on the resistance to all kinds of drugs in tissue culture had the same experience long ago that some of those characters are expressed only in culture, not in the intact plant. If you again make cell suspensions from the plant then that expression returns. This is a very general phenomenon, that some characters are expressed in cultured cells and not in the plant, and it has nothing to do with cytoplasm—it may be a nuclear gene which is expressed under certain conditions, not under others (e.g. Widholm 1980).

Cocking: Concerning what Professor Zenk has just said about the developmental pathway consideration, there is another point in relation to changes in variation associated with the developmental pathway from cell to plant. One of the questions we should address is that many of us want to avoid such variation, we want to have a true clone of our system. Therefore I would like to hear a little more about how the pathway of development of the plant from the tissue culture system is known to affect the variation encountered in the regenerated plant. There is some suggestive evidence in the plant tissue culture literature that if the regeneration is via somatic embryogenesis initiated at an early stage from the cells in culture, this minimizes most aspects of somaclonal variation. Some workers are deliberately trying to use this in their work. In our own work on rice, which is a diploid, with possibly minimal tendencies in this connection, we are trying to induce somatic embryogenesis at as early a stage as possible. Are there any current explanations of this? It might explain some of the points raised by Professor Zenk; he might be able to maintain the disparity in the levels of nicotine production, if he altered the pathway of plant regeneration.

Scowcroft: The consensus about the level of variation from different culture type systems, either protoplast, callus, or somatic embryogenesis, is that variability tends to be reduced in plants derived by somatic embryogenesis. But even so, there are still genotypic differences in the level of variability, within a species.

Fowler: Can I come back to Professor Cocking's point about integration of the whole plant. We did a similar experiment to Professor Zenk but with opium poppy cell lines. We cultured cells from high-yielding poppy plants, which, as appears to be typical, lost the ability to form the major opiate alkaloids in culture, although the presence of certain precursors was demonstrated. When we regenerated whole plants from these cell cultures, the plants synthesized a full profile of alkaloids with a high yield. This brings us back to previous comments that the key problem is the lack of understanding of the physiological integration of the whole plant and its development, and the controls involved in the expression of many of these secondary pathways. A culture is essentially a very aberrant environment for a cell to be in, a point which we need to consider very carefully.

Harms: People have tried to regenerate plants by different routes but I don't know of any system that is experimentally clean and where you have the opportunity of choosing either somatic embryogenesis or morphogenesis and in which you can be sure that your morphogenesis is not an aberrant somatic embryogenesis system. There is a lot of speculation as to the potential effect of those different morphogenetic routes on the level and nature of genetic variability that you get out of the system but there is no clear experimental evidence that I am aware of. Even if you had experimental evidence from one system, would it really allow you to extrapolate to the situation in a different plant? I think we will have that enigmatic or speculative situation for some time without an answer.

Cocking: But this really highlights the point that it is basic studies on plant development that are needed, whereby, whatever the species, you can trigger off the development, either through a callus to organogenesis to a plant, or get that cell to establish polarity and go on to form an embryoid. We need comparative studies in that vein in a number of different species and only then will we begin to get some of the answers that we need, and through getting those answers to know how to control this system—whether to generate more variation or to generate less. I think it's very interesting, not being involved in secondary product work myself directly but looking over the fence and visiting various companies, some here in Japan, to see that there is a gradual swing towards almost tame plants in culture with hairy roots growing all over the place. If you ask, 'Where are the cells?', they take you to see the hairy roots, and I think that's an indication of our lack of knowledge in that we have to go to such hairy root cultures now to make the advances.

Withers: I agree with Dr Harms that we don't have completely clean experimental systems, but there are situations where one could compare a permissive type of regeneration (i.e. embryogenesis) with an induced type (e.g. shoot regeneration) and within that we could look at, for example, cell cycle control and see if that might give a clue to the difference between the two situations. The generation of instability might then be related to the late replication of DNA, for example. So, going to the cell and the control of nuclear events between and during cell division might help find an answer.

Fowler: To go back to the organization/developmental question: in microbial fermentations, for instance, there are many examples where developed fungal mycelial cultures synthesize the desired secondary metabolites, as opposed to single cells. This is a very similar situation to that seen for many higher plant cell cultures. There are lots of precedents in fungal fermentations in the way that those are handled compared to some of our plant systems – they are looking at organized structures. In consequence, those of us working in cell culture should recognize that precedents already exist in fungal physiology.

Rhodes: I would like to take up Dr Cocking's point about organized structures. The instability of many cell culture systems in producing secondary metabolites that Professor Zenk demonstrated several years ago led our group to investigate hairy roots as a source of such compounds. I would not suggest that hairy roots are necessarily always the best system for the production of secondary metabolites, but they are stable in terms of chromosome number and the level of product formation over many cycles, and their use avoids these problems of variation that have been described. It is interesting that the production of these secondary products is closely related to tissue organization. If auxin treatments are applied to hairy root cultures of the solanaceous plants e.g. *Nicotiana, Datura*, there is a rapid loss of root integrity and of the ability of the culture to make the secondary product. If transformed roots of these species are cultured in the presence of auxin, callus formation is stimulated and a suspension develops which does not produce alkaloids. If these transformed cell suspensions are transferred back into hormone-free media, root regeneration occurs and the productive capacity is restored. This is an interesting system which illustrates our lack of knowledge of the regulation of gene expression in relation to secondary product formation.

Potrykus: As someone who has no experience with the production of secondary metabolites, I think you are asking too much from cell cultures if you want them to compete with the highly organized differentiated system which has evolved over millions of years in plants. Many plants have very special organs, with very special cells in which you get high production of a specific compound. This cannot be mimicked by treating cells with known growth regulators. To reproduce the high levels of production that can occur in these specialized cells and organized plants may require the specific differentiation of these cells and this might be quite difficult to achieve.   I would also like to add a comment concerning the desirability of studies on plant development. I fully support this but I think it is naive to believe that within the next 20 years we will completely understand plant development at the molecular level. I have been rather disappointed with the results of all the many growth regulator studies. I think that all we are doing at the moment is playing around at the level of 'pattern realization', without addressing the issue of 'pattern determination'. I think that on this specific level we lack really good ideas and good experimental systems.

Yamada: At what level would you like to see these studies initiated, at the gene level or the chromosome level?

Potrykus: In everything I know about tissue culture, we are using a capacity of the cell which has already been established and we don't know how to establish this capacity. There is the complex phenomenon of 'competence' which is a nice word but which only describes our ignorance. We don't understand how a cell becomes competent to respond to the triggers we are using, to produce the required secondary metabolite or to make an embryo instead of a callus. This 'competence' requires a lot more study.

Rhodes: In the area of secondary products, it is possible to isolate from mixed cultures variant cells with very high rates of secondary metabolite production, but it is often difficult to stabilize these variants as highly productive cell lines.

Yamada: We can establish stable cell lines, it is difficult but possible.

Fowler: I understand from work you have done and others at Mitsui, that if one takes a protoplast and clones it, this yields a more stable cell line than if one clones a cell with an intact cell wall—is that correct?

Yamada: No. If we select a cell line with high metabolite productivity from small cell aggregates (Yamamoto et al 1982), I believe that we get a more stable cell line than from protoplast-derived clones. I don't know if this is due to the multicellular organization of cell aggregates or to some kind of mutation that may occur during cell division of protoplasts.

Fowler: Presumably, we have returned to the concept of stabilization through some degree of multicellular organization.

Takebe: Dr Scowcroft, is there any means of controlling the frequency of somaclonal variation in tissue culture either by manipulating the culture conditions or giving some additives?

Scowcroft: No, the only answer is to avoid prolonged callus culture phases.

Potrykus: An important observation is that some plant species vary hardly at all and others are very variable. Even genotypes within one species respond very differently. If one wishes to avoid variation one should choose a genotype which doesn't vary. However, it would be worth studying why one genotype varies but another does not.

Scowcroft: But you don't know that until after you have done the experiment.

Potrykus: Of course, but you don't know anything until then.

References

Breiman A, Felsenburg T, Galun E 1987 *Nor* loci analysis in progenies of plants regrown from the scutellar callus of bread wheat. Theor Appl Genet 73:827–831

Tabata M 1978 In: Frontiers of plant tissue culture. University of Calgary Press, Calgary p 213–222

Widholm M 1980 Differential expression of aminoacid biosynthesis control isozymes in plants and cultured cells. In: Sala F et al (eds) Plant cell cultures: results and perspectives. Elsevier, Amsterdam p157–159

Yamamoto Y, Mizuguchi R, Yamada Y 1982 Selection of a high and stable pigment-producing strain in cultured *Euphorbia millii* cells. Theor Appl Genet 61:113–116

Applications of cell and tissue culture in tree improvement

Don J. Durzan

Department of Environmental Horticulture, University of California, Davis, California, 95616, USA

Abstract. Trees of value in agriculture and forestry require different strategies in tree improvement. Forest trees continue to require domestication because many of the superior trees have been harvested and the remainder are disappearing from natural stands at an alarming rate. By contrast, fruit, nut and ornamental trees, being already domesticated and cloned, could benefit from wider genetic variation. Global problems of increasing human populations, diminishing resources, both renewable and non-renewable, and crises of various types, e.g. food, energy, environmental, demand improved strategies for woody perennials. These should maintain existing genetic gains by clonal propagation, introduce new variation using classical and modern biotechnologies, and reduce risks by the management of germplasm through cell, life, production and utilization cycles.

Genetic gains can now be captured by somatic embryogenesis and polyembryogenesis in cell suspension culture. Cells can be cryopreserved until genotype x environment x maturity state interactions are worked out for precise environmental adaptation. The recent availability of morphogenic protoplasts creates opportunities to construct novel hybrids through cell fusion and genetic engineering. In forestry, the production of artificial seeds throughout the year provides a complementary technology to reduce risks in the practice of seed orchards where seed production is erratic and uncertain. In fruit crops, the rescue of somatic embryos based on controlled crosses contribute to wider genetic variation. Cell lines from a variety of explant sources throughout the life cycle now enable the study of somatic parthenocarpy, and enhance opportunities for the production of secondary products *in vitro* (directed totipotency).

1988 Applications of plant cell and tissue culture. Wiley, Chichester (Ciba Foundation Symposium 137) p 36–58

The aim of this review is to outline appropriate biotechnologies for the improvement of woody perennials, namely, fruit, forest and ornamental trees. Emphasis will be placed on recent developments in protoplast, cell and tissue culture, and on the initial applications of genetic engineering to morphogenic cells of commercially important woody perennials. The application of biotechnologies to woody perennials follows naturally from recent developments in plant sciences described in this volume, but uses different strategies from those applied to annual crops. In this context, the spectrum of innovations which will be needed and the genetic gains to be captured, are illustrated. Genetic gains represent responses to selection which include: trait heritability, selection differentials, correlations for gene expression in juvenile and mature trees, and stategies to shorten the generation interval (Timmis et al 1987). Selected cases of successful application of biotechnology to woody perennials are described. Emphasis is placed on somatic embryogenesis and polyembryogenesis in previously highly recalcitrant species and on trends in tree improvement programmes.

Problems and opportunities

The increasing global demand for high quality trees reflects the issues and crises created by diminishing renewable resources and increasing human populations. Unfortunately, there is still a range of constraints that create barriers to future tree improvement and to the productivity needed to meet this demand (Durzan 1980, 1982, 1985a, Haissig et al 1987). Issues that are pertinent to the productivities of woody perennials are demands for energy, fuel, structural materials, transportation, chemical feedstocks, environmental impact assessment, and food. There is also a need to protect current resources against genetic erosion, insects, disease, fire and other natural disasters, pollution (e.g. acid deposition) and desertification. A spectrum of innovations are needed to overcome these barriers to greater productivity and economic well-being.

One approach to tree improvement portrays problems and opportunities as a series of linked cycles (Fig. 1). Cycles are based on an hierarchy of complexity involving the elements of tree improvement that lead to commercialization of the resource. Vegetative propagation is currently a pivotal technology in this approach, bringing large trees with their long, complex life cycles into the laboratory in manageable ways for tree improvement programmes. It also plays a role in productivity cycles involving clonal forests, orchards and commercial products. Opportunities for tree improvement extend from the molecular level to the utilization cycle.

A basic stategy in tree improvement is to capture the existing genetic gains by domestication and clonal propagation, then to introduce new genetic variation through the application of classical and modern biotechnologies, and finally to reduce risks in the overall series of cell, life, production and utilization cycles. Genetic attributes that are commonly sought in tree improvement include rapid growth rate, pioneering traits for environmentally harsh areas, resistance to insects and disease, plant-machine compatibility, tree shape and form, fibre quality, fruit quality, and ease of vegetative propagation. Taken singly or together, these traits, when introduced into populations and into production cycles, are called genetic gains.

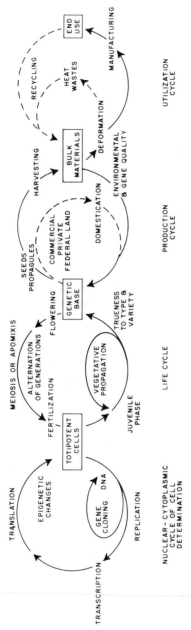

FIG. 1. Tree improvement of woody perennials via a series of production cycles. Production is cyclical, the output from one cycle becoming the input for the next, leading to an end product which ultimately depends on existing renewable germplasm resources. It is depicted arbitrarily as four stages: cell, life cycle, production, product use. Biotechnology enables the cycles to be shortened and altered to make formation of the end product (boxes) more efficient. (1) At the cellular level, genetic gains for tree improvement are captured by cell and tissue culture. Recombinant DNA methods permit the cloning of elite genes and their introduction into the germplasm through totipotent cells. (2) The life cycle begins after fertilization or by cell and tissue culture from totipotent cells. Vegetative propagation may be used to maintain traits such as hybrid vigour and disease resistance often lost by traditional breeding methods. (3) In the production cycle, the perennial habit and improved culture techniques can be exploited to minimize alternate bearing and unpredictability of the crop. Rapid wood production in juvenile trees is needed in forestry but not when the trees are grown for their seed, fruit or nut. (4) In the utilization cycle, quality attributes can be identified and introduced back into the breeding programme to improve further the efficiency of the utilization cycle. End uses of bulk materials are a function of supply and demand.

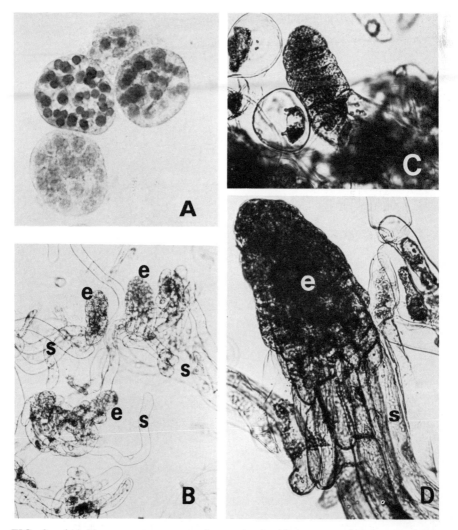

FIG. 2. (A) Norway spruce protoplast stained with acetocarmine to reveal free nuclear stage about 7–8 days after agarose embedding (Boulay MH, Durzan DJ, unpublished). (B) Polyembryogenic cell suspension culture of Norway spruce showing proembryos (e) and suspensor cells (s) obtained from embryonal-suspensor masses recovered from ripe seeds. (C) Emergence of tiered proembryo from an embryonal-suspensor mass of Norway spruce stained with acetocarmine. (D) Proembryo development in cell suspension cultures grown at one rpm in a nipple flask. The proembryo (e) stains red with acetocarmine and the suspensor (s) stains blue with Evan's blue.

Capturing genetic gains

The capturing of genetic gains can be portrayed as an equation with four components: heritability of the trait sought; a selection differential to enrich

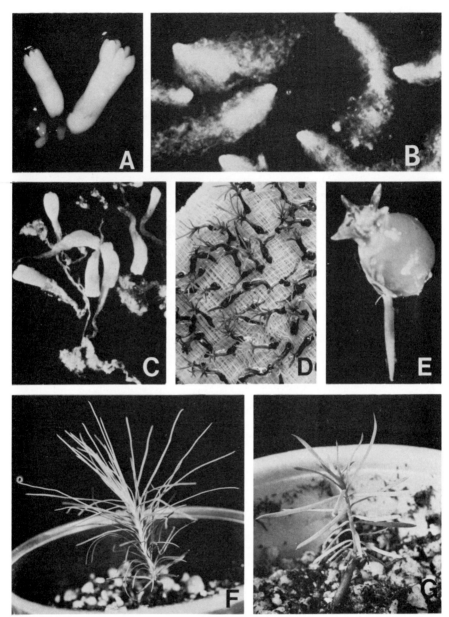

FIG. 3. (A) Rescued sugar pine embryos from a five-year-old seed representing a controlled cross between two blister rust-resistant parents (cf Gupta & Durzan 1986). The embryonal-suspensor mass at the base of the largest embryo, isolated with the rescued embryos, continues to generate more embryos by somatic polyembryogenesis. (B) Individual development of somatic embryos of loblolly pine with elongated suspensors in a cell suspension culture of the embryonal-suspensor mass. Individual embryo

that trait in the population; a statistical correlation for trait performance to reduce the progeny testing time with long-lived trees; and methods to save time in terms of trait introduction through breeding cycles and crop rotation (Durzan 1985 a,b). In this context, we have to consider the liabilities and variations introduced by clonal methods. For clonal systems, these concerns have been reviewed by Timmis et al (1987) and Kester (1983).

Historical events and strategies leading to improvement of forest trees are: unconstrained forest harvesting, natural stand management and managed plantations followed by tree breeding. The latter is now being reinforced by vegetative propagation (rooting, cell and tissue culture) (Hall 1980) and genetic engineering (Durzan 1980). Currently, much of the improved germ-plasm comes from long-range breeding programmes and seed orchards. This approach remains costly, unwieldy and unpredictable for many commercially valuable species. Fortunately, some of our best remaining trees can now be selected, rooted, micropropagated, micrografted, rejuvenated, and cloned by somatic embryogenesis and polyembryogenesis. For horticulture, priorities and cautions have been outlined by Zimmerman et al (1986).

These advances may soon be obtained with the aid of cell suspension cultures and with protoplasts (e.g. Gupta & Durzan 1987a,b, Figs. 2 and 3). Cryopreservation of embryogenic cell lines enables cost-effective mainte-nance of germplasm until clonal progeny have been evaluated under field conditions (Gupta et al 1987). Field performance involves the interaction between genes and the environment, and between a clone and its maturity state. By extending this approach, embryonic–juvenile–mature correlations may be established for the reliable expression of elite traits. Where possible, mature trees may be rejuvenated for propagation purposes to capture gains of proven elite specimens (Fig. 4) (Gupta & Durzan 1987c). In the commercial realm, clonal methods are now emerging that are suitable for gymnosperms (conifers) (Bonga & Durzan 1982, 1987) and angiosperms (*Prunus* sp.) (Zim-merman et al 1986, Fig. 5). The next step is to add further value to elite clonal material by introducing new useful genetic variation.

---

development was encouraged by addition of abscisic acid to culture medium. Abscisic acid inhibits the cleavage process. (C) Population of individual somatic embryos of Douglas fir with cotyledon primordia, suspensors and remnants of embryonal suspen-sor masses used to initiate the suspension cultures. (D) Population of germinated somatic embryos with cotyledons and, in some cases, newly emerging leader shoots. Roots are covered with dark coleorhizae. Each propagule is about 1.5 cm in length. (E) Alginate-encapsulated Norway spruce somatic embryo that has germinated at 20° C after 2 to 3 weeks in the gel. (F) Loblolly pine in soil 2 to 3 months after germination of the somatic embryo. Growth occurred in continuous white light at 23 ± 1° C. (G) Douglas fir in soil 5 to 6 weeks after germination of the somatic embryo. Growth occurred in continuous white light at 23 ± 1° C.

42

New genetic variation

Prospects for the introduction of new genetic variation have been helped by the recent availability of embryogenic cell lines for protoplast production. The early stages of somatic embryogenesis have been obtained from conifer protoplasts (Gupta & Durzan 1987a). Furthermore, a foreign gene (fire-fly luciferase) has been introduced into morphogenic protoplasts by electroporation (Gupta et al 1988, Table 1). Protoplasts with regenerated cell walls show transient expression of this gene, detectable when the substrate for the gene product is added. The recovery of plants from morphogenic protoplasts sets the stage for protoplast fusion and for the introduction of useful complex traits already inherent in somatic cells of existing populations. Useful genetic traits add further weight to the heritability factor, and through their properties the most appropriate and applicable selection differential can be identified. The introduction of new genetic variation into proven, *mature* trees is also now possible. Micropropagated shoots from mature Douglas fir have been genetically transformed by *Agrobacterium tumefaciens* bearing genes for opine production and kanamycin tolerance (Dandekar et al 1987). Evidence for gene integration was obtained by Southern blot analysis.

With mature fruit trees, the fate of some cell lines, e.g. cherry petiole cells in suspension culture, can be directed to somatic parthenocarpy (Fig. 5). It is too early to evaluate the potential of such 'directed' cell totipotency in *Prunus* sp. for the introduction of new genes. The reduction of long-lived trees to cells and the recovery of developmental processes based on rejuvenation or senescence now allows complex biological processes to be studied and manipulated in the laboratory.

FIG. 4. Micropropagation of Douglas fir using explants from 60-year-old trees (cf Gupta & Durzan 1987c). Sprouting of axillary buds (arrows) from stem tips with half-pruned-back needles on DCR medium (20 days). (B) Elongation of buds on 0.5 DCR medium (30 days). Left, plagiotropic growth of shoots from explants collected from sprouts of basal axillary buds. Right, orthotropic growth of shoots from explants from branches with an upright habit in the lower crown of trees. (C) Multiple shoot bud formation on DCR medium with N^6-benzyladenine (0.5 mg/litre) from nodal segments excised from *in vitro* shoots after 25 days of the fourth subculture. (D) Elongation of buds on 0.5 DCR medium after 35 days. (E) Adventitious buds with new leaves develop on needle surface (ns) on DCR medium with BAP (0.5 mg/litre). Needles were excised from sprouting buds of shoots after fourth subculture (12 weeks). Bar represents 0.8 mm. (F) Rooted plantlet from 60-year-old tree on 0.5 DCR medium after 10 weeks. (G) Plantlets growing in pots containing sterile peatmoss, vermiculite and perlite (1:2:1 volume) (four months).

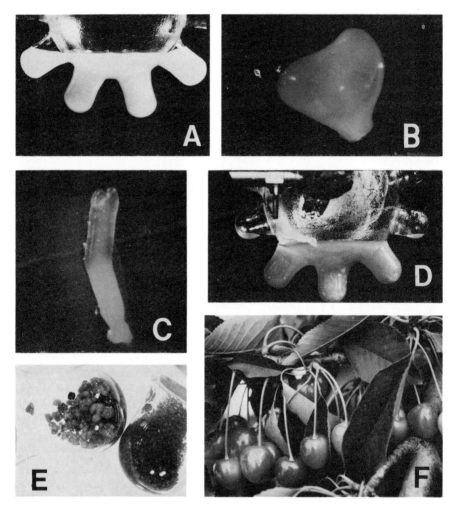

FIG. 5.   Somatic embryogenesis and somatic parthenocarpy in cell suspension cultures of *Prunus* sp. (A) Cell suspension culture of cherry rootstock cells obtained from callus of petioles from a 29-year-old tree. Cells are grown in darkness by weekly subculture in a one litre nipple flask. (B) Somatic embryo from cell suspension culture after 3 to 4 weeks of development (Bhansali R, Durzan DJ, unpublished). Isolated embryos were cultured on agar plates in light to promote later stages of development. (C) Elongated somatic cherry embryo derived from cell suspension culture after two months. Now grown in light it is about 8 mm long. (D) When the cherry rootstock cell suspension culture in (A) is grown in light, cells in suspension turn a bright red after 7 days of culture, in response to changes in the medium (Peng C, Durzan D, unpublished). (E) The cell suspension in (A) is divided in half and grown in light. Under the influence of specific substrates, cells in the nipple flasks develop mesocarp-like red nodules (right), those in the other flask remain green and become green nodules (left). The development of colour and fragrances contrasts with somatic embryogenesis and represents attributes associated with fruit ripening. (F) A totipotent cell suspension can

TABLE 1   Expression of luciferase activity in viable (fluorescent detectable reactivity) and morphogenic (acetocarmine reactivity) protoplasts 36 hours after electroporation of the foreign gene into freshly prepared protoplasts

	$LU/1 \times 10^6$ protoplasts			
	Douglas fir Without PEG	With PEG	Loblolly pine Without PEG	With PEG
No DNA	0	0	0	0
With DNA (luc)	52	105	103	196

Light unit (LU) = light flashes produced by 1 pg of firefly tail enzyme extract in 1 min. PEG, polyethylene glycol; luc, luciferase.

Limits of recent applications and innovations

This section relates mainly to protoplast, cell and tissue culture strategies and to certain future trends. We must recognize the positive impact of vegetative propagation and micropropagation on tree improvement over the past decade (Bonga & Durzan 1982, 1987). Today, somatic embryogenesis, polyembryogenesis and the introduction of foreign traits using protoplasts, fusion techniques, microinjection and *Agrobacterium* vectors create new opportunities, but these are expectations at the research and laboratory level. These methods are currently being applied to woody perennials in laboratories and tested in field studies throughout the world. One trend is to enhance tree improvement in tropical areas where resources are rapidly being eroded and endangered.

It now appears that all conifers and probably most gymnosperms can be cloned by somatic embryogenesis and polyembryogenesis (Durzan & Gupta 1988, Gupta & Durzan 1986, 1987a,b). This includes the recovery of embryos from protoplasts (Gupta & Durzan 1987a). Figs. 2 and 3 illustrate the various stages of embryo recovery. It is important to note that the embryos are not derived from a callus, but from pro-embryonal cells bearing a characteristic acetocarmine-reactive neocytoplasm. The neocytoplasm is considered to be a product of the fusion of the male and female nucleus at fertilization. We have found that the neocytoplasm reacts strongly with acetocarmine. This reactivity enables us to assay populations of cells for their morphogenic potential before experimental studies and scale-up or pilot runs.

---

be derived from petiole callus from a fruit-bearing tree, directed towards either somatic embryos or attributes of the developing mesocarp. This is in some ways a restoration from petiole cells of the cherry fruit from the tree but without the stone layer and epidermis.

In somatic polyembryogenesis, clonal products are reconstituted via a true-to-type basal plan of development, which includes a free-nuclear stage. The embryogenic process clones the new generation rather than the mother tree as was done in haploid cultures of larch (Nagmani & Bonga 1985) or in diploid cultures of the nucellus of *Citrus* (Spiegel-Roy & Vardi 1984). Moreover, the zygotic system for conifers is of the cleavage polyembryonic type. Where zygotic embryogenesis is not normally polyembryonic, as in Douglas fir (Durzan & Gupta 1988), polyembryogenesis can be induced in suspension cultures of somatic cells and the development of individual embryos encouraged by the addition of abscisic acid. The positive effect of abscisic acid was illustrated with suspension cultures of Norway spruce (Boulay et al 1988, Durzan 1987) and of loblolly pine (P.K. Gupta, D.J. Durzan, unpublished). However, abscisic acid leads to an arrest of embryonic processes that may be difficult to remove when conversion to plantlets is attempted. More study is needed on factors that enhance the conversion of embryos treated with abscisic acid, on seeding performance and physiological pre-conditioning, and on lethal aspects of the somatic polyembryonic process, particularly in haploid cell lines (Dogra 1967). Control over these factors is a prerequisite to the requirement for accurate and precise environmental adaptability under field conditions. Somatic embryos from the new generation can now be recovered from the cotyledons or other parts of the embryo or seedling by reproductive regeneration (Krogstrup 1984, Hakman et al 1985). This process needs to be extended to mature trees.

Populations of clonal materials in the form of propagules or artificial seeds will require considerable field testing. In some cases, testing could take at least nine years, as in Douglas fir. Since it would not be cost-effective to subculture cells every 10 days for nine years, we have explored the cryopreservation of embryonal cells in liquid nitrogen ($-196\,°C$). Thawed cells of Norway spruce and loblolly pine have regenerated plantlets (Gupta et al 1987). While cells have not yet been stored for a long time, their dense cytoplasmic and spore-like characteristics make them ideal material for cryopreservation because such cells are less readily disrupted by the freezing and thawing process. We do not yet know the effect of long term cryopreservation on embryo recovery and field performance.

When somatic embryos are encapsulated as artificial seeds, we have conceptually an ideal system to complement current seed and breeding orchard technology. The expectation is that elite material may be produced clonally throughout the year, independently of damage and losses due to insects, climate, natural disasters, pollution or poor seed years. However, in reality, current artificial seed technology is far from being representative of a true seed. Hence, alternative and improved embryo conveyance mechanisms are now the subject of interest in several laboratories.

By applying diagnostic methods developed for somatic embryogenesis in cell suspensions, we can now isolate stocks of elite embryonal cells for the preparation of morphogenic protoplasts (Gupta & Durzan 1987a, Rao & Akins 1985). These may someday serve as carriers of new genetic information. Currently, they enable studies of gene expression during development. The most promising genetic modification of gymnosperm protoplasts appears to be with the gene for the *Bacillus thuringiensis* toxin to provide resistance to lepidopteran insect attack. Several methods are now available to insert and integrate foreign genes (Fraley et al 1986).

For most woody perennials, we have yet to explore the potential of protoplast fusions resulting in the rapid creation of new hybrids with proven, more useful, naturally occurring traits. For field verification, the very complex long-lived trees will require decades of testing before the biotechnologies really can be accepted as reliable, cost-effective, and free of adverse effects. The trend that emerges is one where appropriate technology, field establishment, rapid screening and quality assurance of clonal products, are major concerns for future progress.

Our experience with fruit trees opens the door for a greater exploitation of apomixis and the life cycle (Fig. 5.) With a 28 to 30-year-old cherry rootstock, and with rescued embryos of peach, nectarine, plum and apricot, we can now induce *in vitro* repetitive multiple embryogenesis in embryos that would normally abort (R Bhansali, J Driver, DJ Durzan, unpublished). The rescued embryos represent crosses made by breeders for commercially important traits, such as early ripening and fruit quality. Rescued embryos of *Prunus* have also yielded embryo-producing cell suspension cultures. Invariably, the cells competent to produce embryos react diagnostically with acetocarmine, as do the gymnosperm cells. Hence, the same technologies of protoplast fusion and foreign gene integration are being explored with *Prunus*. Only time will tell if such methods will contribute to the improvement of fruit trees.

Of special interest, however, is the fact that petiole cells of a mature cherry tree can not only be induced to produce low frequency somatic embryos from cell suspensions, i.e. be rejuvenated to an embryonic state, but can also be directed to behave like cells of the developing and ripening mesocarp (DJ Durzan, C Peng, unpublished). In the latter, cells produce red pigments and in some cases, fragrances of cherry, which do not occur in the orchards at Davis. This process, which yields nodules of cherry-like flesh, we tentatively call somatic parthenocarpy. It enables the study, separation, control and exploitation of partial reactions in the developmental processes associated with fruit production and ripening. It also shows that one can reconstruct the components of the fruit and seed as discrete developmental alternatives from the same cell suspension.

Future developments with woody perennials may also exploit the directed totipotency of somatic cells to manufacture secondary products (Durzan 1988), i.e. the cells may be grown to produce the needed tissue masses, then the secondary products induced internally or released as volatiles, as has been demonstrated by Fujita (this volume).

Outlook

In forestry, the commercialization of clonal products tends to be inhibited by the comparatively low value (0.1 to 0.5 dollars US) of the product. Certified value must somehow be added to gain profitability, e.g. disease-resistant clones containing *B. thuringiensis* genes to make the product more attractive in the market place. By contrast, trees of high ornamental and/or food-producing value, but in less demand numerically, usually have greater individual market value by factors of 10 to 100. These provide interesting short-term commercial opportunities and germplasm for multiplication and improvement using vegetative propagation and recombinant DNA methods. Nevertheless, for woody perennials, the current lack of availability of novel new products, and the long testing and take-over time for any new product of biotechnology remain problematical.

The near future will still be dominated by research. We can anticipate a confluence of current biotechnologies with other technologies that ultimately will add value to potential new products and increase the efficiency of production. Now that several commercial laboratories have the capacity to produce millions of woody plant propagules annually, we can expect that clonal methods will be combined with risk-reducing technologies that lead to improved quality and process control by mechanization and computerization and to the certification of stock.

References

Bonga J, Durzan DJ (eds) 1982 Tissue Culture in Forestry. Martinus Nijhoff/Dr. W. Junk, The Hague, The Netherlands

Bonga JM, Durzan DJ 1987 Cell and tissue culture in forestry. 1. General principles and biotechnology. 2. Scientific principles and methods. 3. Case histories: Gymnosperms, angiosperms, and palms. Martinus Nijhoff, Dordrecht

Boulay MH, Gupta PK, Krogstrup P, Durzan DJ 1988 Development of somatic embryos from cell suspension cultures of Norway spruce (*Picea abies* Karst). Plant Cell Rep, in press

Brink RA 1962 Phase change in higher plants and somatic cell heredity. Q Rev Biol 37:1–22

Dandekar AM, Gupta PK, Durzan DJ, Knauf V 1987 Genetic transformation and foreign gene expression in micropropagated Douglas-fir (*Pseudotsuga menzieii*). Bio/Technology 5:587–590

Dogra PD 1967 Seed sterility and disturbances in embryogeny in conifers with particular references to seed testing and breeding in Pinaceae. Stud For Suec 45:1–97

Durzan DJ 1980 Progress and promise in forest genetics. In: Paper sciences and Technology: the Cutting Edge, 50th Anniversary, May 9–10, 1979, The Institute of Paper Chemistry, Appleton, Wisconsin, p 31–60

Durzan DJ 1982 Cell and tissue culture in forest industry. In: Bonga J, Durzan DJ (eds) Tissue Culture in Forestry. Martinus Nijhoff/Dr. W. Junk, The Hague, The Netherlands, p 36–71

Durzan DJ 1985a Biotechnology and the cell cultures of woody perennials. For Chron 61:439–447

Durzan DJ 1985b Tissue culture and improvement of woody perennials: an overview. In: Plant Tissue Culture: application to agriculture and forestry. Symposium on plant cell and tissue culture. September 9–14, 1984, University of Tennessee, Knoxville, Tennessee, p 233–256

Durzan DJ 1987 Improved somatic embryo recovery. Bio/Technology 5:636–637

Durzan DJ 1988 Nitrogenous extractives extracellular to the lignocellulosic cell wall. In: Rowe JW (ed) Natural products extraneous to the lignocellulosic cell wall of woody plants. Springer-Verlag, New York, in press

Durzan DJ, Gupta PK 1988 Somatic embryogenesis and polyembryogenesis in Douglas-fir cell suspensions. Plant Sci 52:229–235

Fraley RT, Rogers SG, Horsch RB 1986 Genetic transformation in higher plants. CRC Crit Rev Plant Sci 4:1–46

Fujita Y 1988 Industrial production of shikonin and berberine. In: Applications of plant cell and tissue culture. Wiley, Chichester (Ciba Found Symp 137) p 228–238

Gupta PK, Durzan DJ 1986 Somatic polyembryogenesis from callus of mature sugar pine embryos. Bio/Technology 4:643–645

Gupta PK, Durzan DJ 1987a Somatic embryos from protoplasts of loblolly pine proembryonal cells. Bio/Technology 5:710–712

Gupta PK, Durzan DJ 1987b Biotechnology of somatic polyembryogenesis and plantlet regeneration of loblolly pine. Bio/Technology 5:147–151

Gupta PK, Durzan DJ 1987c Micropropagation and phase specificity in mature, elite Douglas-fir. J Am Soc Hort Sci 112:969–971

Gupta PK, Durzan DJ, Finkle BJ 1987 Somatic polyembryogenesis in embryogenic cell masses of *Picea abies* and *Pinus taeda* after thawing from liquid nitrogen. Can J For Res 17:1130–1134

Gupta PK, Dandekar AM, Durzan DJ 1988 Somatic proembryo formation and luciferase gene expression in Douglas-fir loblolly pine protoplasts (submitted)

Hakman I, Fowke LC, Von Arnold S, Eriksson T 1985 The development of somatic embryos in tissue cultures initiated from immature embryos of *Picea abies* (Norway Spruce). Plant Sci Lett 38:53–59

Haissig BE, Nelson ND, Kidd GH 1987 Trends in the use of tissue culture in forest improvement. Bio/Technology 5:52–59

Hall FK 1980 Biology and genetics. In: Paper science and technology: the cutting edge, 50th Anniversary, May 9–10, 1979, The Institute of Paper Chemistry, Appleton, Wisconsin p 15–17

Kester D 1983 The clone in horticulture. Hortscience 18:831–837

Krogstrup P 1984 Micropropagation of conifers. (In Danish). PhD thesis, Dept Horticulture, Royal Veterinary and Agriculture Univ. Copenhagen, Denmark

Nagmani R, Bonga JM 1985 Embryogenesis in subcultured callus of *Larix decidua*. Can J For Res 15:1008–1091

Rao PS, Akins O 1985 Plant regeneration from embryonic suspension-derived protoplasts of sandalwood (*Santalum album*). Protoplasma 124:80–86

Speigel-Roy P, Vardi A 1984 Tropical and subtropical fruits: citrus. In: Ammirato PV et al (eds) Handbook of plant cell culture, MacMillan, New York III:355–372

Timmis R, Abo El-Nil M, Stonecypher RW 1987 Potential genetic gains through tissue culture. In: Bonga JM, Durzan DJ (eds) Cell and tissue culture in forestry, Martinus Nijhoff, Dordrecht, vol. 1, 198–215

Zimmerman RH, Griesbach RJ, Hammerschlag FD, Lawson RM (eds) 1986 Tissue culture as a plant production system for horticultural crops. Martinus Nijhoff, Dordrecht

DISCUSSION

Hall: I am interested in the development of the tumours on those little shoots. Were you looking for opines as well? Were these wild-type Ti plasmids?

Durzan: The transformed Douglas fir cells displayed autotrophic growth in culture, synthesis of octopine, the presence of foreign DNA sequences and expression of a foreign gene. The plasmids were of two types: K12X562E and K12X167, both derivatives of the wild-type plasmid, pTiA6. The plasmids contained a chimaeric foreign gene encoding resistance to the antibiotic kanamycin (Dandekar et al 1987).

Hall: So the shoots that you were getting would probably not form roots, nevertheless there was opine production. Were there opines present in the regenerated differentiated leaves?

Durzan: We haven't looked at that. We simply had a source of micropropagated material from these older trees and this was an opportunistic observation.

Hall: A very exciting observation.

Durzan: The cell suspension culture work with embryogenic cell lines of conifers using the acetocarmine and Evans blue stains is interesting. We can take a cell suspension culture and predict in advance how many embryos we should get back from that particular cell suspension culture. We can also use these stains to search the life cycle of a tree for cells that have embryogenic potential.

Withers: I am only familiar with Evan's blue as a mortality stain (Withers 1978), is there a difference in the procedure to use it in the way you do?

Durzan: No, the procedure is the same. The elongated suspensor cells that are associated with embryonic development normally push the pro-embryo into the nutrient reserve. Those elongated cells are alive, but their membranes are leaky, so that the Evan's blue, which is normally associated with mortality, is able to enter those suspensor cells.

It's also interesting that if you take the free nuclear stage, either as a protoplast or in cell suspension cultures, not all of the nuclei stain equally—some nuclei appear to take up more acetocarmine and Evan's blue than others. This means that at the free nuclear stage, the differentiation of the embryo may already be determined to a certain extent, before the nuclei migrate. When we cryopreserve those cells in liquid nitrogen, all the elongated cells that normally stain blue are killed by the freezing and thawing process; one is left only with the densely cytoplasmic embryonic cells which stain red with acetocarmine. As these develop and regenerate back into plants, one can see suspensor cells growing off of them, so we know that the blue staining property comes from cells that started off with the ability to stain red with acetocarmine.

Potrykus: So you are using acetocarmine as a vital stain?

Durzan: No, just as an interesting and very simple marker. Within three minutes you know the result with sample subpopulations of cells.

Potrykus: This is an aqueous solution of acetocarmine?

Durzan: Yes.

Potrykus: When you regenerated shoots from this tumour transformed material, were you able to find opines?

Durzan: We found compounds that co-chromatograph with opines but we could not get a Southern blot from shoots to show the presence of the gene that confers antibiotic resistance. We only tried once, twelve shoots about 5–6 cm high weren't enough to do the assay properly. We did find opines and obtained Southern blots to show that we could pick up the genes in the DNA from Douglas fir tumour cells.

Harms: You said they were growing on 'high kanamycin', what concentration did you use?

Durzan: Twenty to 50 milligrams per litre for shoots and 50–200 milligrams per litre for tumours on agar plates.

Harms: But they did not root at this concentration?

Durzan: We didn't look at rooting: we were pleased to be able to regenerate shoots from tumours.

Barz: Could you clarify how you direct your process from embryogenesis to the production of the red pigment in the cherry system?

Durzan: It's an interesting and unusual algorithm that involves a number of factors acting on cells initially grown in the dark: one is light, another is a change of nutrients in the medium, a third involves a shift in plant growth regulators.

Barz: That is highly comparable to what Professor Komamine has found with the carrot system: embryogenesis versus the production of anthocyanins. In your case, Professor Komamine, is it correct that the differentiation process only depends on 2,4-D and the number of cells per aggregate?

Komamine: By removing 2,4-D, we induce embryogenesis and anthocyanin synthesis in carrot suspension cultures, but these two inductions did not occur in the same cells. In some rather lighter cells, we can induce the synthesis of anthocyanin and in heavier cells we can induce embryogenesis. Could you induce synthesis of anthocyanin and embryogenesis in the same cells?

Durzan: We haven't tried that at this stage. We simply have the two alternative sets of factors: when we set up one situation we get low frequency embryogenesis: the embryogenesis is not spectacular, those cells want to produce roots more than embryos. If we stopped that process and exposed the cell suspension culture to an alternative set of factors, including light, nearly all of the cells turned red but we haven't obtained definitive numbers for this.

Barz: Did you look for any other compound in your red cells which is characteristic for the cherry fruit?

Durzan: No we did not. However, although cherries don't usually produce much fragrance, when we added an amino acid supplement to the cell suspension culture it gave off a set of unusual fragrances.

Potrykus: Is this a reproducible process or was it an isolated case?

Durzan: We started in 1982, we've repeated this every year since then and now we have a Gene Expression Laboratory in Albany, California trying to repeat our observations independently. We appreciate that comment because our cells are derived from woody perennials; every year there is a new set of leaves and you have to ask whether or not other explant sources would give the same results, and you have to repeat this year after year. The red coloration part of it and the embryogenesis look pretty good.

Tabata: Could you tell me more about the behaviour of free nuclei during embryogenesis? Does the embryo come from one of these nuclei?

Durzan: Yes. Just before the pro-embryo forms one sees a population of free nuclei (depending on the species). These nuclei migrate and then one set of nuclei (or a nucleus) gives rise to the elongated suspensor, the other(s) forms the pro-embryonal head. That is one of the differences between the embryogenic process in gymnosperms and some angiosperms. Many of the angiosperms form embryos by internal cell divisions, but the gymnosperms seem to go through this free nuclear stage. This encourages us to believe that we have a true-to-type embryogenic process in these cells.

Tabata: So one of these nuclei migrates to the pole of the cell?

Durzan: That's right, and as the nucleus migrates, it starts to release a cytoplasm around it and that cytoplasm is very reactive with acetocarmine.

Tabata: In the case of angiosperm protoplasts, we have worked with fennel protoplasts (Miura & Tabata 1986). The protoplast expands first to become a large ellipsoidal cell, then the nucleus migrates to the extreme end, or pole, where it divides to form a pro-embryo.

Zenk: That gene transfer now works in conifers is really important: I think you could tackle some of the major problems in forestry, for example, manipulating the content of cellulose versus lignin. You might also be able to introduce syringye units and convert a softwood into a hardwood.

Durzan: The firefly genes were inserted by electroporation and it was only a transient expression, but we were gratified to see foreign gene expression by the use of that method. So all we are saying at the moment is that there are a lot of opportunities for future research.

Barz: Have you examined how plants regenerated by this process perform under natural conditions, for example, with regard to their resistance to pathogens?

Durzan: The testing stage has not yet reached a significant point. A number of interesting problems have arisen, especially in the species that show this cleavage polyembryogenesis. One adds abscisic acid to stop this cleavage process, but in some cases too much abscisic acid leads to an embryo arrest which is very hard to break later on. In cases where one does not use abscisic acid to prevent cleavage, the problem has been to recover enough clonal germplasm for testing in a range of natural habitats. Weyerhaeuser in the United States has already repeated a good part of the work on the embryogenic process. I understand that they now have nearly a thousand Norway spruce generated by somatic polyembryogenesis and in a few years they will be able to answer your question. Initial testing involves taking embryos that come through the cell suspension culture process and seeing how they perform under field conditions. That testing is going to take a very long time. Douglas fir may require nine years of testing before anybody would dare put that material out into commercial orchards.

Scowcroft: What is your assessment of the time scale for obtaining good somatic embryogenesis from mature tissue? That is the essence of what you are trying to achieve: going back to seedlings or to seed material doesn't really do much more than the forestry industry is already doing, because those populations of trees that provide your seed source are genetically very heterogeneous.

Durzan: That scenario is true only in part. Several US timber companies have access to seed orchards based on genetically improved trees, i.e. in an industrial context, the resource base may be less heterogeneous than the question implies. Useful strategies, already employed with seed and seedling material in a typical tree improvement programme, have been described by Timmis et al (1987). Nevertheless, your question indicates the direction in which one should be looking. We need somatic embryogenesis or some improved cloning procedure with mature proven trees. I think it will take 5–10 years before that happens, and that's being optimistic. In my experience, embryogenesis starting with invigorated cells from mature trees only takes us to this pro-embryo stage. Moreover, the embryogenic process for some conifers is very long: from pollination to obtaining a ripe seed in sugar pine takes three years. I think that perhaps by the end of another decade, one should be able to get reliable somatic embryogenesis from invigorated tissues from mature trees. Part of this relates to this acetocarmine-reactive factor: one of the characteristics of these embryogenic cells is that when they grow in that embryonal-suspensor mass (this is not a callus), it produces a gel. This consists of a mucilaginous material which, when spread over these explants, contains a factor(s) that contribute to the early stages of organized development in the explants.

Scowcroft: You say it will be another decade before reliable embryogenesis of mature material will be practical. Because of the 10–15 years required for field evaluation, it seems unlikely that micropropagation of forest trees will be a commercial reality for at least two decades.

Durzan: But some companies already have seed orchards and breedings orchards where they have improved material coming along.

Scowcroft: This is where my next question comes and I thought you might have addressed that when you were looking at juvenile versus mature correlations. To what extent do you think there may be the capability to use, for example, restriction fragment length polymorphisms correlated with characteristics such as lignification or the length of fibre in mature trees, to enable a more efficient screening of seedling nurseries for candidates for commercial production.

Durzan: That's a possibility but one of the problems with gymnosperm cells is that they have a tremendous amount of DNA, they have more DNA than most plant cells. Currently available DNA technology—unless it has a little more specificity— may not help. We know very little about DNA sequence polymorphisms within populations and how to use this information to make inferences about population structure or juvenile–mature correlations.

Potrykus: If I understood you correctly, there is so far no definite proof that genes have been transferred into shoots from fir trees. One should therefore still be a little cautious about predicting too far ahead with these possibilities. It seems fantastic that this embryogenesis from protoplasts is so identical to the normal embryogenic process in a fir. Can you exclude the possibility that during your protoplast isolation you included, from meristematic material, multinucleate protoplasts which had arisen by spontaneous fusion. Have you followed the development of a single protoplast to a multinucleate protoplast in an individual example, or is this interpretation of stages you find later?

Durzan: We have not done this as rigorously as we would like. There is spontaneous fusion, one does see protoplasts with four or five nuclei from this spontaneous fusion, but the zygote of Norway spruce has a stage with 32 nuclei. We have seen close to this number of nuclei in protoplasts, but I've never seen spontaneous fusion to that extent. While we have not followed the fate of these multinucleate protoplasts or cells, an interesting opportunity exists to study this free nuclear stage for its genetic and developmental implications.

Galun: There are other groups working on micropropagation of pine trees, for example, in the laboratory of Aitken Christie in New Zealand. What is the main difference between the methodology which they use and the propagation of pine trees from mature tissue?

Durzan: It relates to explant source. Propagation of pine trees or any other conifer from mature tissues usually involves rejuvenation of tissues followed by the rooting of cuttings or micropropagation. As yet, reliable somatic embryogenesis is not possible with explants from mature trees. In somatic embryogenesis with conifers, one can use a combination of plant growth regulators and induce embryo formation from the epidermal cells or sub-epidermal cells of a cotyledon placed on an agar plate (Durzan & Gupta 1988). In contrast, with somatic polyembryogenesis, we have been able to culture the embryonal-suspensor mass that is an early product of fertilization and scale that up. In the seed and from that mass, normally only one embryo is produced, the rest are suppressed by inhibitors or some other phenomenon. By isolating and culturing the proliferating embryonal-suspensor mass of cells, we were able to release this inhibition so that these cells continued to repetitively reconstitute embryos for over two years in cell suspension culture. From this experience it is not impossible that somatic embryogenesis will be obtained in the future from explants from mature trees.

Riley: I'm interested in the control of somatic embryogenesis, because of its relationship to apomixis. In apomixis, a somatic embryo is formed quite often in the nucellar tissue. There is evidence of the genetic control of apomixis and therefore of the genetic control of somatic embryogenesis. It appears to me that undifferentiated tissue often moves automatically to the development of polarity, and therefore to the creation of somatic embryos. Do we understand why it is that in some plant cellular systems somatic embryos are formed relatively frequently, but in others this process is absent or rare? Is it an automatic process that undifferentiated tissue becomes polarized?

Potrykus: I would be surprised if anybody knew anything about the control—you can observe both kinds of development in identical material. Dr Harada knows far more about this than I, but from our own experience with a single population of microspores, with slight modifications of environmental conditions which don't control but allow development, you can have perfect embryogenesis or culture formation. I think it is as I suggested before, pattern determination is brought about by endogenous triggers which are at present obscure.

Professor Durzan, do I remember correctly that you have reported here that you could regenerate somatic embryos and plants from protoplasts of Douglas fir?

Durzan: No, not plants but the early stages of somatic embryogenesis have been observed, that is, we can recover the pro-embryo developing with the suspensor cells. That is as far as we have got at present, this is all very recent work.

Potrykus: But it is continuing so there is hope that this will lead to plants.

Durzan: That's right, but such studies require continuity in research support.

Fowler: Once you have established somatic embryogenesis, can you synchronize the process?

Durzan: I am trying to improve synchrony at the moment. The maintenance of the cells in suspension culture is a cyclical process. The cells are dividing, they go through a free nuclear stage and then develop into a pro-embryo and then there is a cleavage phenomenon. This sequence occurs in darkness and forms a continuous cycle that can be maintained for a long time. The cells will stay in that particular cycle and so there is initially an asynchronous population of cells. When we want to initiate the completion of embryogenesis, we add abscisic acid. Pro-embryos that are poised for later embryo development then develop under the influence of light, so that the later stages appear to be more synchronized than the starting population. This activity initiates a second cycle leading to fully developed somatic embryos. It would be interesting to try to synchronize that very first cycle and see how synchrony of the cell cycle relates to the free nuclear stage and subsequent development.

Fowler: That is an important point. The impression one gets from people in micropropagation is that somatic embryogenesis needs to be synchronized and to produce large numbers of viable embryos for this approach to be commercially attractive. I get the impression from our engineering colleagues that it is not the technical problem of selecting out the embryos for potting and then planting which constitutes the difficulty, but rather our lack of biological control and ability to achieve a high level of synchrony in embryo formation. In our early work, we could certainly get quite good synchrony of carrot cultures by controlling the dissolved oxygen tension and the phosphate level.

Durzan: The tolerance of this system to changes in conditions appears to be very low—one has to subculture every ten days, keep them in darkness, and so on. After three years we have not obtained complete synchrony but the development of the embryos seems to be more in phase. However, other problems arise that affect the populations of embryonal and suspensor cells in liquid culture.

Fowler: We used to achieve about 70% synchrony over about a five hour period. You are quite right that the tolerance was limited in terms of timing of the application of the control signals, we couldn't vary them much at all.

Cocking: Just a comment, Chairman, about possible genetic manipulation approaches to rejuvenation of mature tree species that we are beginning to explore in relation to fusion. There is some interest in exploring whether fusions between juvenile material and mature material at the protoplast level, perhaps only involving cytoplasmic interactions, might lead to novel rejuvenation of the mature system. The theory is difficult to explain in detail but at least the possibility exists of experimentation in that direction now that at least a few tree species are beginning to be developed at the protoplast level. Propagation from mature elite trees is often desired but usually only juvenile trees can be vegetatively propagated. Transfer of juvenile cytoplasm via protoplast fusion might enable ready propagation from mature trees.

Chaleff: We heard from Bill Scowcroft about the genetic variability associated with cell culture. By contrast, Professor Durzan's system, in which somatic embryogenesis is used to propagate tree species, would depend on genetic fidelity. Given the long maturation time of tree species, have you yet been able to evaluate the variation in your populations of regenerated plants?

Durzan: We have not been able to evaluate the variability as much as we would like. There is some indication that the rate of embryo development and the morphogenesis are true to type. In terms of control factors or the genotypes of these plants/trees it is hard to say. Conifers are polyembryonic. Several embryos can be formed from a single egg cell or from other gametophytic or sporophytic cells. Also, several eggs may develop from one megaspore and each may be fertilized separately or activated by parthenogenesis (simple polyembryogenesis). The result is embryos of a fraternal type. Moreover, more than one plant may arise by division or cleavage of a single embryo, to give genetically identical progeny (cleavage polyembryogenesis). One has to be able to distinguish these factors. In gymnosperm embryogenesis, there is another phenomenon whereby in this free nuclear stage sometimes a nucleus comes from the mother tree and confounds the process. In the literature (Singh 1978), these are called relic nuclei and they introduce additional genetic variability. Whatever happens in early seed development—be there relic nuclei or identical or paternal types of embryos—usually only one embryo comes out of that embryonal-suspensor mass. So by taking that embryonal-suspensor mass, we are getting all different types of possibilities, at least from the hypothetical viewpoint. On that basis, if one could recover efficiently all the embryos coming from that mass, there should be some genetic variability. This genetic variability would not have translated into developmental aberrations at this point, perhaps because of our use of genetically improved seeds and limited population sampling. Genetic variability is difficult to evaluate: we have found that to make sure that you get the early fidelity in development, the system has to be kept in complete darkness—once you give a little bit of light to that embryonal suspensory mass, there is a lot of aberrant development. When you exceed these light thresholds there are a lot of aberrations. We have to control the composition of the culture media, the temperature, the light, the rate of subculture and so forth, then the development at least is very true to type. Yet that could represent quite a heterogeneous population of genotypes.

Scowcroft: I think that research in New Zealand on radiata pine is probably further ahead than that of anybody else in terms of field assessment of the trees derived from micropropagation. My understanding is that the plants derived by micropropagation of radiata pine do tend to breed true-to-type. There is some variability, however. This variability is not considered a serious problem because the relative comparison is the level of variability seen in a seed-produced seedling nursery where plants are derived from outcrossed seed which is genetically quite variable.

Riley: It is desirable to generate heterogeneity in the population anyway, for simple biological protection.

Scowcroft: Yes, but in that particular case it would be preferable to generate controlled heterogeneity using genetically different clones from micropropagation.

Riley: Of course, but a little uncontrolled heterogeneity is not going to make very much difference.

Durzan: One has to be careful when comparing different systems. You talk about trees that are cloned by micropropagation but you have to keep in mind the explant source, the cells that give rise to those individuals come usually from cotyledons or they are induced from another somatic cell that is somewhat along the line in development. There are genetic changes that can occur in these particular cells and I'm not sure how fair it is to compare plants that have been generated by micropropagation to those generated by somatic embryogenesis in terms of some still undefined product specification.

Scowcroft: I was thinking of a comparison between somatic embryogenesis and seed produced material, not micropropagation.

Durzan: I don't know of anybody in New Zealand who has obtained plants by somatic embryogenesis. It is important not to confuse true somatic embryogenesis with organogenesis or micropropagation.

Tabata: It is very difficult to evaluate whether all the plants commonly propagated from somatic embryos are genetically uniform or not. We have some experience with regenerated plants of *Coptis japonica*, *Angelica acutiloba*, and *Foeniculum vulgare* (which are used for medicinal purposes) by somatic embryogenesis. We have conducted field tests for a number of years and compared them morphologically and chemically (Nakagawa et al 1982, Miura et al 1987). The coefficients of variation of both morphological and chemical characters are very narrow compared with control plants, seed-propagated from heterozygous plants. I think the plants propagated from somatic embryos are extremely useful as raw materials for producing crude medicinal materials. We can only talk about statistical differences. All the chromosome numbers are the same as those in the parental plant but I don't know about the minor genetic changes that might arise during embryogenesis or even before.

References

Dandekar AM, Gupta PK, Durzan DJ, Knauf V 1987 Genetic transformation and foreign gene expression in micropropagated Douglas-fir (*Pseudotsuga menzieii*). Bio/Technology 5:587–590

Durzan DJ, Gupta PK 1988 Somatic embryogenesis and polyembryogenesis in conifers. Adv Biotechnol Processes 9:53–81

Miura Y, Tabata M 1986 Direct somatic embryogenesis from protoplasts of *Foeniculum vulgare*. Plant Cell Rep 5:310–313

Miura Y, Fukui H, Tabata M 1987 Clonal propagation of chemically uniform fennel plants through somatic embryoids. Planta Med *1987*:92–94

Miura Y, Fukui H, Tabata M 1987 Reduced homogeneity of *Angelica acutiloba* plants propagated clonally through somatic embryoids. Planta Med *1987*, in press

Nakagawa K, Miura Y, Fukui H, Tabata M 1982 Clonal propagation of medicinal plants through the induction of somatic embryogenesis from the cultured cells. In: Fujiwara A (ed) Plant tissue culture Maruzen Co, Tokyo p 701–702

Timmis R et al 1987 Potential gain through tissue culture. In: Bonga JM Durzan DJ (eds) Cell and tissue culture in forestry. Martinus Nijhoff, Dordrecht vol 1:198–215

Withers LA 1978 The freeze preservation of synchronously dividing cultured cells of *Acer Pseudoplatanus* L. Cryobiology 15:87–92

The induction of embryogenesis in *Nicotiana* immature pollen in culture

Hiroshi Harada, Masaharu Kyo* and Jun Imamura**

Institute of Biological Sciences, University of Tsukuba, Tsukba-shi, Ibaraki, 305 Japan

Abstract. Immature pollen grains consitute unique experimental material for the study of embryogenic processes because they are uniform in size and shape in a given species and can easily be isolated. They are also useful in the production of haploid and homozygotic plants for breeding and studies on genetics. However, basic information on pollen embryogenesis is limited and the mechanism of the acquisition of embryogenic capacity is still unclear. This is mainly due to the fact that the number of immature pollen grains which undergo the process of embryogenesis is not high enough to allow physiological and biochemical analysis.

Therefore, with the aim of increasing the frequency of embryogenesis, we conducted a series of experiments by both anther and isolated pollen culture on various factors that affect *Nicotiana* pollen embryogenesis. The results enabled us to increase significantly the rate of embryogenesis and to direct the development of immature pollen grains to either normal maturation or embryogenesis.

We also attempted to define any biochemical changes which may be related to the induction of pollen embryogenesis by applying this experimental system. We found some significant changes in protein phosphorylation which seem to be related to the initial stage of pollen embryogenesis.

1988 Applications of plant cell and tissue culture. Wiley, Chichester (Ciba Foundation Symposium 137) p 59–74

The elucidation of the mechanisms of cell, tissue and adventitious embryo and organ differentiation is one of the most important and intriguing research subjects in plant physiology. Several different types of studies on differentiation have been conducted by a large number of researchers, but many questions remain unanswered. It was reported for the first time by Guha and Maheshwari (1964, 1966) that immature pollen cells could develop to haploid embryos. Since then, pollen embryogenesis has raised a number of intriguing questions in developmental physiology and has aroused much interest on the part of plant breeders, geneticists and physiologists as an experimental system and as a means to provide useful tools for plant breeding. Up to now, more than 500 papers have been published on this subject and pollen

---

Present address: * Faculty of Agriculture, Kagawa University, Mikicho, Kagawa, Japan. ** Plantech Research Institute, Kamoshida, Yokohama, Japan.

embryogenesis has been reported in nearly 200 species. However, information about the physiological and biochemical changes occurring in the initial period of pollen embryogenesis is scarce and the basic mechanisms of the acquisition of embryogenic capacity are little known. Pollen grains are fairly uniform in size and shape within a single species or cultivar and, in many cases, they can easily be isolated by gently pressing anthers in a liquid medium, giving a homogeneous cell suspension. Thus, they offer unique experimental material for the study of cell differentiation.

Many investigators have attempted to analyse the process of asexual embryogenesis of plants, but biochemical investigation of *in vitro* embryo differentiation has been quite limited. This is probably due to three major experimental difficulties: 1) the number of embryogenic cells in a given tissue is very small; 2) cells at an early stage of embryogenesis cannot easily be detected; and 3) the synchronous induction of embryogenesis in a large number of cells is difficult to achieve. To solve these problems, adequate experimental systems must be established. For this purpose, we have selected two plant materials, namely immature pollen cells of *Nicotiana rustica*, and *N. tabacum* as both intact anthers and isolated pollen cells. Until recently, the frequency with which immature pollen grains gave rise to haploid plants was very low, despite much effort to increase it. In order to improve the frequency of pollen embryogenesis, we have conducted a series of experiments on various physical, chemical and physiological factors that affect *Nicotiana* pollen embryogenesis, using both anther and isolated pollen culture techniques. Secondly, we used our experimental system to investigate which biochemical changes may be related to the induction of pollen embryogenesis. We found some significant changes in protein phosphorylation which seem to correlate with the initial stage of pollen embryogenesis. This paper summarizes some of the results we have obtained during the past seven years (Imamura & Harada 1980 a,b,c,1981, Kyo & Harada 1985, 1986).

Materials and methods

Anther culture

The anthers of *N. tabacum* cv. Samsun grown in a greenhouse were used for all experiments except those employing reductants. Flower buds at the stage of pollen mitosis and the early binucleate stage were harvested when they reached the appropriate corolla length (13–16 mm). Anthers of *N. tabacum* cv. MC were used for experiments with reducing agents. Sterilized anthers were placed on 0.8% agar medium containing Nitsch's medium elements (Nitsch & Norreel 1974), 2% sucrose and 0.4% activated charcoal. Petri dishes (6 cm in diameter) containing 30 anthers each were subjected to a photoperiod of 16 hours light and eight hours darkness at 28 °C.

Treatments with reduced atmospheric pressure and an anaerobic environment were as follows. Anthers in open Petri dishes were put aseptically into a glass desiccator under reduced pressure for different time intervals. Air was aspirated by a vacuum pump at 12 mmHg. For anaerobic treatments, either N_2 or a mixture of N_2 and O_2 (21% to 2.5% O_2) was passed aseptically for various time periods through the desiccator at 70 ml/min at atmospheric pressure. Each experiment was repeated 3–10 times.

Isolated pollen culture

Flower buds of *N. rustica* and *N. tabacum* cv. Samsun that had corollas of 7–8 and 18–22 mm, respectively, were harvested. Pollen grains in these buds were mostly at the mid-binucleate stage. After sterilization, the anthers were gently pressed with a pestle in 0.4 M mannitol, then filtered through a nylon mesh with a pore size of 53 μm to obtain free immature pollen grains. Isolated pollen was cultured in Petri dishes and various treatments were applied. Different media consisting basically of the macroelements of Murashige & Skoog's medium (1962) were used, as shown in Figs. 2 & 3. Experimental details have been published elsewhere (Kyo & Harada 1985,1986)

Results

Anther culture

Effects of reduced atmospheric pressure. Treatment with reduced atmospheric pressure increased both the frequency of anthers producing plantlets and the average number of plantlets per anther (Table 1). Significant results

TABLE 1 Effects of reduced atmospheric pressure on pollen embryogenesis in N. tabacum cv. Samsun 40 days after the beginning of culture

		Duration of treatment				
	Control	10 min	20 min	60 min	90 min	24 hours
Number of anthers cultured	33	37	37	36	33	37
Anthers with plantlets (%)	6	65	54	50	49	0
Total number of plantlets produced	57	447	640	603	343	0
Average number of plantlets per anther	2	12	17	17	10	0

TABLE 2   Effects of N₂ on pollen embryogenesis of N. tabacum cv. Samsun 40 days after the beginning of culture

	Duration of treatment (minutes)			
	Control	*15*	*30*	*60*
Number of anthers cultured	32	32	32	31
Anthers with plantlets (%)	34	44	72	71
Total number of plantlets produced	290	225	807	770
Average number of plantlets per anther	9	7	25	25

were obtained with 10, 20 and 60 minute treatments. However, 24 hour treatment completely inhibited plantlet formation.

Effects of anaerobic environment.   To investigate whether the stimulatory effect of reduced atmospheric pressure on pollen embryogenesis was due mainly to reduced air pressure itself or to decreased partial pressure of O_2, anthers were kept under a 100% N_2 stream at one atmosphere for 15, 30 and 60 minutes (Table 2). The 30 and 60 minute treatments promoted the produc-

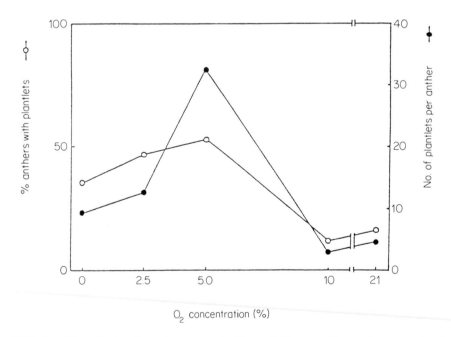

FIG. 1.   The effects of varying concentrations of O_2 on pollen embryogenesis of *Nicotiana tabacum* cv. Samsun. One hour treatments, 34 anthers each. Zero on the abscissa indicates 100% N_2.

TABLE 3 Effects of abscisic acid (10^{-5}M) treatment on pollen embryogenesis of N. tabacum cv. Samsun

	Duration of treatment (days)			
	Control	1	2	4
Number of anthers cultured	32	30	29	22
Anthers with plantlets (%)	63	67	93	36
Total number of plantlets produced	117	398	199	102
Average number of plantlets per anther	4	13	7	5

tion of plantlets. Treatment with N_2 containing 0, 2.5 and 5% O_2 increased the frequency of anthers generating plantlets and the average number of plantlets per anther (Fig. 1). These data indicate that *quasi*-anaerobic conditions (2.5 or 5% O_2) are more effective than purely anaerobic treatment (100% N_2).

Effects of abscisic acid.   The three day treatment with abscisic acid (10^{-5}M) increased the frequency of anthers producing plantlets as well as the number of plantlets formed per anther (Table 3). The one day treatment showed the highest average number of plantlets produced per anther.

Effects of reducing agents.   Continuous treatment with ascorbic acid at concentrations between 0.1 µg/ml and 100 µg/ml and mercaptoethanol at concentrations between 10^{-6} M and 10^{-3} M increased the frequency of anthers that gave rise to plantlets and the average number of plantlets per anther. Dithiothreitol also showed a stimulatory effect on embryo formation from pollen grains.

Isolated pollen culture

The general scheme of our method with *N. rustica* is shown in Fig. 2. A crucial feature of this method is to give no carbon source to isolated pollen for an initial period of culture (starvation period, 1st culture), then to supply an adequate carbon source, e.g. galactose. When no starvation period was given, isolated pollen grains scarcely divided but accumulated starch grains, as seen in normally developing pollen. Therefore, we presume that the process of dedifferentiation and the acquisition of embryogenic capacity in cultured pollen occur during the starvation period. The optimum period of starvation (1st culture) is temperature dependent, being three days at 30 °C but 7–9 days at 15 °C.

The fractionation of pollen by Percoll discontinuous gradient centrifugation was used to select a homogeneous population of pollen grains at specific developmental stages. Embryogenic cells were derived from pollen grains at the mid-binucleate stage.

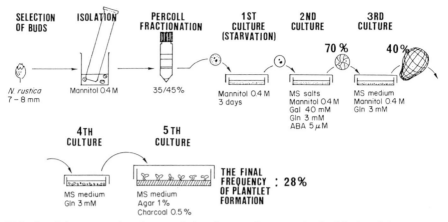

FIG. 2. Diagrammatic scheme of the direct culture method of isolated immature pollen grains of *Nicotiana rustica*. Ten days after the start of the 2nd culture, the cell population consisting of approximately 70% dividing pollen was transferred to the 3rd medium, where about 40% of dividing pollen developed into embryos (or large embryogenic calli). Mannit, mannitol; gal, galactose; gln, glutamine; ABA, abscisic acid.

For the 2nd culture, during which pollen begins to divide, the presence of galactose in the medium was important. Among different plant growth substances tested, abscisic acid at concentrations between 10^{-5} and 10^{-7} M was effective in stimulating pollen cell division in the 2nd culture. However, abscisic acid inhibited the further development of divided pollen to form embryos or large calli.

We could also induce pollen embryogenesis in *N. tabacum* cv. Samsun with minor modification of culture medium (see Fig. 3). In this species, the frequency of embryo or callus formation from divided pollen was lower than that in *N. rustica*, but the frequency of induction of pollen cell division was high (40–70%) using our method. The pollen maturation process in *N. tabacum* cv. Samsun could be achieved *in vitro* by giving glutamine during the 1st culture, if the starting material was pollen at the same developmental stage as that used for the induction of embryogenic cell division. The proportion of cultured immature pollen that attained full maturation *in vitro* was about 90%, of which 50% (maximum) germinated when transferred into 0.3 M sucrose solution containing 2 mM boric acid (Fig. 3).

Changes in phosphorylated protein pattern

We have conducted a preliminary experiment with the aim of finding any changes in the two-dimensional electrophoretic patterns of newly phosphorylated proteins from immature pollen of *N. tabacum* cv. Samsun cultured under various conditions. When the mid-binucleate pollen dedifferentiated in

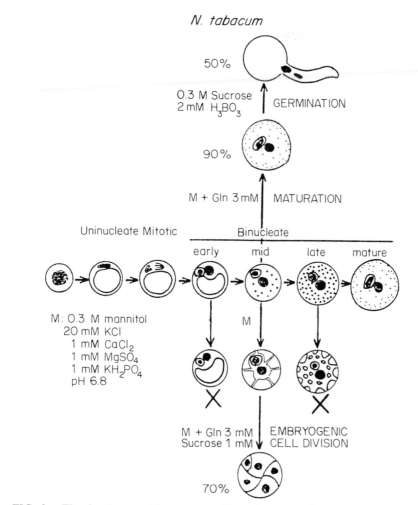

FIG. 3. The developmental processes of *N. tabacum* cv. Samsun pollen *in vitro* and *in vivo*. In intact anthers, pollen develops by the normal maturation process (left to right); after the mitotic stage, starch grains accumulate, increasing the density of pollen. By selecting flower buds with a certain corolla length and then applying Percoll fractionation to isolated pollen, it was possible to obtain a fairly homogeneous pollen population at each developmental stage. Using the mid-binucleate pollen population which was characterized by the absence of vacuoles and the presence of a few starch grains, we were able to induce dichotomously at a high frequency both maturation (above) and embryogenic cell division (below) *in vitro*.

basal medium without glutamine, phosphorylation of the proteins named a-d was clearly detected (Fig. 4). When pollen grains matured in medium with glutamine, phosphorylation of proteins named e-i was evident but there was no apparent phosphorylation of proteins a-d. The phosphorylation of proteins

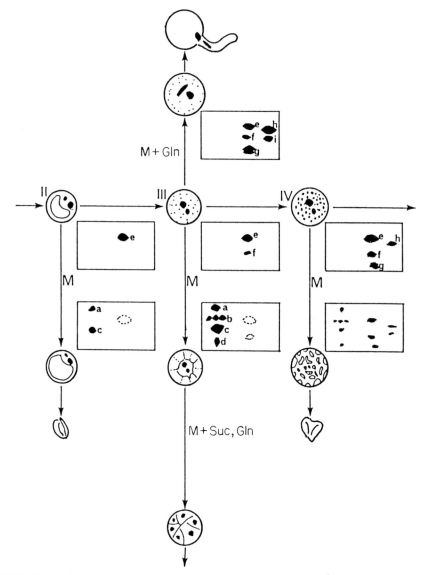

FIG. 4. Variations in protein phosphorylation during embryogenic induction and maturation of *Nicotiana tabacum* immature pollen in culture. The cells were radio-labelled with ^{32}P for 12–48 hours, the proteins resolved by 2-dimensional gel electrophoresis and autoradiographed.

a-d was partially or completely suppressed in pollen grains at stages other than mid-binucleate which did not transform to embryogenic cells in medium without glutamine. When chemical compounds which promote (EDTA) or inhibit (6-benzyladenine) pollen dedifferentiation were added to the medium, the electrophoretic pattern of phosphorylated proteins was changed.

Discussion and conclusions

We described earlier some of the difficulties which hinder the progress of physiological and biochemical studies on *in vitro* somatic embryogenesis of higher plants. Practically no experimental system has so far been available for studies on the physiological processes of differentiation starting from truly free single cells of higher plants. In order to solve some of these problems, we established appropriate experimental systems and conducted a series of investigations related to different factors that affect plant cell differentiation employing *N. rustica* and *N. tabacum* cv. Samsun and MC.

Tobacco pollen embryogenesis may be divided into two physiologically distinct stages, namely induction of the embryogenic state and embryo formation. The induction process requires approximately three to eight days, depending on culture conditions such as temperature. The direction of pollen development can be influenced during this period, for example, by switching its course away from gametophyte development to sporophyte development. The effects of different chemical and physical stresses on immature pollen were examined with the aim of elucidating the fundamental mechanism of pollen embryogenesis and also enhancing the frequency of haploid plant production for practical purposes.

Except for certain plants belonging to a few genera, such as *Nicotiana* and *Datura*, the frequency of pollen embryogenesis in culture has been very low. Our studies on *Nicotiana* anthers and isolated pollen indicate that similar treatments may increase the frequency of embryo formation in other plant species in which pollen embryogenesis has so far been considered difficult or impossible (Wei et al 1986). Even in tobacco, it has been difficult to obtain a consistently high yield of pollen embryo formation.

Our findings will also facilitate future physiological and biochemical studies on pollen embryogenesis. Further investigation is required to answer some of the basic questions — Why and how do these methods stimulate pollen embryogenesis? Does each treatment exert its action through different mechanisms or are common biochemical processes involved?

To study the mode of action of the stimulating factors and to analyse the biochemical processes occurring during the induction stage of pollen embryogenesis, we have been working on direct culture techniques of isolated pollen leading to haploid plant production at a high frequency. We have obtained some significant results, for example, we reported (Imamura et al 1982) that pollen embryogenesis can be induced directly from isolated pollen without a prior culture period or prior treatment of buds or anthers. Our method offers significant advantages in studying the mechanism of pollen embryogenesis in comparison to the use of anther culture; for example, it eliminates unknown effects of anther tissues. Cultured pollen seems to undergo the processes of dedifferentiation and redifferentiation to form embryos. Functional and morphological differences between mature and dedifferentiated pollen induced *in vitro* can be clearly observed. The former possesses developed cytoplasm and germination ability, the latter shows characteristic cytoplasm which remains in the centre of the cell, and possesses cell division ability. Therefore, factors related to these differences in the cytoplasm and in meiotic versus mitotic ability must also affect the pollen maturation and dedifferentiation processes. We expected that these factors would be some high molecular weight substances, especially proteins. Quantitative differences were noted in total protein content and the rate of protein synthesis: during maturation both of them increased, while during dedifferentiation both of them decreased. We propose that the degradation of proteins and/or the suppression of protein synthesis are necessary conditions for pollen dedifferentiation. However, in spite of many experiments and careful observations, we have not so far been able to detect any protein spot specific to dedifferentiating pollen in two-dimensional gel electrophoretic patterns. We still expect to find certain protein changes which characterize the state and the function of dedifferentiated pollen.

Recently, many biologists have been studying the role of protein phosphorylation in the regulation of cell function in response to environmental changes in eukaryotes. In our isolated pollen culture system of *N. tabacum* and *N. rustica*, the glutamine starvation caused immature pollen cells to deviate from the normal developmental pathway and dedifferentiate. The association of protein phosphorylation with pollen dedifferentiation seemed a possibility. Therefore, we analysed the newly-phosphorylated proteins of cultured pollen and found some significant differences between proteins from pollen during the maturation and dedifferentiation processes. When immature pollen was cultured without glutamine or with glutamine at very low concentrations, which allows the pollen to dedifferentiate, proteins a-d were highly phosphorylated. On the other hand, when the pollen was cultured with glutamine at high concentrations which allows it to mature, the degree of phosphorylation of proteins a-d was very low and that of proteins e-i was high. The intensity of phosphorylation of proteins a-d and the proportion of dedifferentiated pollen cells increased coincidently after 24 hours of culture without glutamine. It is possible, then, that the phosphorylation of proteins a-d and e-i is associated in some way with pollen dedifferentiation and maturation, respectively.

Acknowledgement

Present studies are supported in part by the grants-in-Aid (to H.H.) of the Ministry of Education, Science and Culture, Japan.

References

Guha S, Maheshwari SC 1964 *In vitro* production of embryos from anthers of *Datura*. Nature (Lond) 204:497

Guha S, Maheshwari SC 1966 Cell division and differentiation of embryos in the pollen grains of *Datura* in vitro. Nature (Lond) 212:97–98

Imamura J, Harada H 1980a Studies on the changes in the volume and proliferation rate of cells during embryogenesis of *in vitro* cultured pollen grains of *Nicotiana tabacum* L. Z Pflanzenphysiol 96:261–267

Imamura J, Harada H 1980b Stimulatory effects of reduced atmospheric pressure on pollen embryogenesis. Naturwissenschaften 67:357–358

Imamura J, Harada H 1980c Effects of abscisic acid and water stress on the embryo and plantlet formation in anther culture of *Nicotiana tabacum* cv. Samsun. Z Pflanzenphysiol 100:285–289

Imamura J, Harada H 1981 Stimulation of tobacco pollen embryogenesis by anaerobic treatments. Z Pflanzenphysiol 103:259–263

Imamura J, Okabe E, Kyo M, Harada H 1982 Embryogenesis and plantlet formation through direct culture of isolated pollen of *Nicotiana tabacum* cv. Samsun and *Nicotiana rustica* cv. Rustica. Plant Cell Physiol 23:713–716

Kyo M, Harada H 1985 Studies on conditions for cell division and embryogenesis in isolated pollen culture of *Nicotiana rustica*. Plant Physiol (Bethesda) 79:90–94

Kyo M, Harada H 1986 Control of the developmental pathway of tobacco pollen *in vitro*. Planta (Berl) 168:427–432

Murashige C, Skoog F 1962 A revised medium for rapid growth and bioassay with tobacco tissue culture. Physiol Plant 15:427–497

Nitsch C, Norreel B 1974 La cultures de pollen isolé sur milieu synthétique. C R Hebd Séances Acad Sci Ser D Sci Nat 278:1031–1034

Wei ZM, Kyo M, Harada H 1986 Callus formation and plant regeneration through direct culture of isolated pollen of *Hordeum vulgare* cv. Sabarlis. Theor Appl Genet 72:252–255

DISCUSSION

Potrykus: This is a beautiful experiment with a new system, I think that it is one of the nicest systems available to study the regulation of embryogenesis. It is encouraging to see that phosphorylation of proteins gives clear differences.

Yamada: I would like to ask about the second step in your system: you emphasized that galactose supply is very important after the starvation of sucrose, what does this mean in metabolic terms?

Harada: Galactose and sucrose both work but in order to have a homogeneous embryogenic reaction, galactose is better.

Yamada: But in many culture media for plant cells, galactose is not included.

Harada: During the second culture, pollen cells may metabolize galactose more slowly than sucrose. The gradual digestion of galactose by pollen cells seems to induce physiologically favourable conditions in pollen cells for maintaining high viability and also for homogeneous embryogenesis. The switching of the developmental pathway takes place during the starvation period and cell division occurs later, during the second culture.

Komamine: I have a comment concerning the effect of starvation on somatic embryogenesis. We are studying somatic embryogenesis from single cells in a carrot system. In this system, sucrose starvation has an inhibitory effect on the induction of somatic embryogenesis. Do you think that there is any difference between the process of somatic embryogenesis from single cells and embryogenesis from pollen?

Harada: What do you mean by difference?

Komamine: The molecular mechanism or the biochemical mechanism?

Harada: As we discussed this morning, the question of differentiation is so complicated that it is difficult to propose precise mechanisms, but there must be important differences.

Komamine: In your system, has absciscic acid a positive effect on embryogenesis from pollen?

Harada: It has a positive effect on the induction of embryogenesis, but for the development of the embryo it is inhibitory, as far as our plant material is concerned.

Komamine: But abscisic acid has a negative effect on both the induction and development of somatic embryogenesis, that is the difference.

Potrykus: I think this points again to two pathways of embryogenesis. In one, for example the carrot system, development of embryogenic cells into embryos is inhibited by high auxin concentrations; this inhibition can be released by removing the auxin. In the other pathway, you have to positively induce a change in the development of microspores, for example mesophyll cells from *Brassica*. I think we must make a distinction between these two processes because they are developmentally very different.

Galun: I would like to make a comment from our experience with citrus. A citrus callus can be induced to produce embryos by shifting it to a medium containing galactose—this was done by Kochba et al (1978). More recently, it was found that a much more uniform and synchronous production of embryos can be induced by shifting into a medium containing glycerol (Ben-Hayyim & Neumann 1983). Have you tested the effect of glycerol?

Harada: Not with *N. tabacum*. In the case of *N. rustica*, the frequency of pollen cell division was lower with glycerol than with galactose or sucrose.

Galun: There is a tremendous difference in efficiency, glycerol is much more efficient than galactose.

Potrykus: I wonder whether there is a basic difference between the citrus system and this pollen embryogenesis system. As far as I understand, the citrus system represents a *release* of potential, in pollen embryogenesis you shift the development from 'pollen tube germination' to 'embryogenesis'—would you agree with this?

Galun: I agree that these are two different systems, but both have one common denominator, an abrupt change in differentiation.

Scowcroft: In our *Brassica napus* breeding programme at Calgary we routinely use microspore culture. It is now very efficient and the two most important criteria in determining that efficiency are pre-treatment of the plants that produce the anthers which give rise to the microspores, and the genotype. If you raise plants routinely at low day/night temperatures, you will always get efficient microspore propagation: the medium is almost immaterial, it is the condition of the anthers when you remove the pollen grains from them. There is a large genotypic component involved. Have you had that experience in tobacco?

Galun: Not at all, it was always said that the conditions under which the plants were grown were decisive for the yield of androgenic plants.

Harada: We have noted seasonal changes. During winter time we could obtain the highest frequency of pollen embryogenesis, in summer the frequency is lower. Naturally, these seasonal changes are due to the difference in the physiological state of the mother plants.

Fowler: When you initiate the starvation phase of the pollen culture, you have a phosphate component in the medium; is the concentration of phosphate the same at the end of the starvation period as at the beginning?

Harada: We haven't checked.

Fowler: One of the concerns I have regarding the interpretation of your data relates to some rather different work. A problem of using mannitol, glycerol and galactose as osmotica is that phosphate starvation often occurs, because the inorganic phosphate is metabolized and sequestered into mannitol-1-phosphate, galactose phosphate and so on, rendering it unavailable for use in general metabolism. This may in turn limit growth or any form of development, even though theoretically there is plenty of phosphate in the system. In view of the osmotica used in your experimental system, I wondered if you had noted any changes in free phosphate levels which might provide a simpler explanation of your observations.

Harada: We compared the composition of the medium of the first culture with or without phosphate. We found that the presence of phosphate was slightly inhibitory for the dedifferentiation of pollen cells.

Harms: You indicated that glutamine at one point has a determinative effect on the maturation of the pollen. Is this a specific effect of glutamine or can it be repeated by other amino acids such as proline?

Harada: We have tested several amino acids and found that glutamine was most effective. Alanine and proline were also effective in promoting maturation but the frequency of germination of mature pollen was lower than in the presence of glutamine.

Harms: It appears to me that glutamine has an impact as a triggering compound in a number of experimental systems. If you analyse those, they are not always going in parallel. There would appear to be contradictions but these may reflect differences in regulation of biological events. I'm personally looking forward to more work in those areas to elucidate the specific role, if any, of a single compound in triggering one system one way and another system in another way.

Harada: In lower plants, such as certain algae and mosses, or in spore-forming bacteria, yeast, fungi and slime moulds, deficiency of certain amino acids is known to affect morphogenesis and developmental programmes.

Yamada: Have you ever tried using this embryogenesis system for somatic cells? Does one get plants in the same way, or have you only tried this for haploid cells?

Harada: I don't think that in the case of somatic cells the starvation treatment would have any effect. We haven't tried.

Zenk: You mentioned that you now want to investigate the biochemistry of the system: which way would you like to approach the identification of the functional identity of these proteins? Can you use your system to generate mutants?

Harada: We think that the important event is occurring during the starvation period, and that high molecular weight substances, especially proteins, are responsible for this, so we are looking at the protein level. So far, unfortunately, we cannot find any proteins which are specifically related to the pollen embryogenesis.

Zenk: Is mutation possible?

Harada: I cannot say what may be possible.

Fowler: To return to the basic biochemistry—have you looked at the fate of the glutamine when you add it back to the system? We did this and it was a very salutary exercise. The glutamine was being rapidly broken down to α-oxoglutamic acid and used as a carbon source, the amide nitrogen and the amino nitrogen were excreted. We were quite surprised by this; the pattern of metabolic events indicated that the glutamine molecule was not acting as a trigger or signal system but was being used by the cell purely as a carbon source for anaplerotic reactions. We need to examine at a very early stage the metabolic fate of the glutamine before saying whether or not it is a signal system.

Harada: If we put glutamine in the starvation medium, then we would have to think of that, but we added glutamine only to the second medium.

Fowler: Even so, I still think that it is important to look at the metabolic fate of the glutamine molecule before you can designate this as a trigger system or a straightforward carbon or nitrogen source.

Scowcroft: The same argument applies to the effect of mannitol, and possibly starvation, these are both stress-induced responses, and one of the immediate and general consequences of that stress is elevation of proline level or, in some cases, glycine betaine. It may be useful to ask what is the direct effect of compounds like proline and glycine betaine.

Potrykus: To an outsider, it is always surprising to see that there is such a short window in time when microspores can respond either way. Has anybody tried to culture microspores early in development, say starting from tetrads? or even from sporogenic tissue?

Harada: We have tried and we have only succeeded with pollen grains at the mid-binucleate stage.

Potrykus: Do you have an explanation for why this shift in the normal pathway is possible only during a very short time?

Harada: At the present time I have no satisfactory explanation.

Chaleff: I appreciate what Professor Harada is saying, but I would encourage the exercise of more caution in evaluating the effect of an exogenously added hormone on any biological process. What is ultimately important is the endogenous hormone concentration, which is determined not only by the amount of material added but also by the rate of metabolism and the initial concentration within the cell or plant.

Harada: It also depends on the plant species.

Fowler: It is not just the exogenous hormone concentration. We should also consider the endogenous form of the substance that we have added. There are a number of examples, of which sucrose may be the best known, where an externally added hormone or substrate is extensively modified before it enters a cell through the plasmalemma. The important question then concerns the form in which the exogenously applied hormone is active within the cell. The 'trigger' substance may be quite different from the molecule originally applied.

Takebe: A very simple question: are all the plants you get via pollen embryogenesis haploid?

Harada: So far as we have checked, most of them were haploid; a few seemed to be autodiploid.

Potrykus: You talked about isolated microspore culture in barley, which I find very exciting. You said that you have a ratio of approximately 2:1 albinos :greens, which is a favourable ratio because as far as I know barley is very problematic and produces lots of albinos. Would your system, in which you can isolate single microspores, allow you to study the reason for the production of albinos? For example, whether the green plants are derived from vegetative cells and the albinos come from the generative cells.

Harada: If we had a homogeneous population of microspores of barley which produced only green plants or albino plants then we could study this, but at the microspore stage it is difficult to distinguish the two types.

Potrykus: Did you ever try the microculture system developed by Koop and Schweiger (1985)?

Harada: No, not yet.

Potrykus: This might be a very potent technique by which to study this question.

References

Ben-Hayyim G, Neumann H 1983 Stimulatory effect of glycerol on growth and somatic embryogenesis in *Citrus* callus cultures. Z Pflanzenphysiol 110:331–337

Kochba J, Speigel-Roy P, Neumann H, Saad S 1978 Stimulation of embryogenesis in *Citrus* tissue culture by galactose. Naturwissenschaften 65:261

Koop HU Schweiger HG 1985 Regeneration of plants from individually cultivated protoplasts using an improved microculture system. J Plant Physiol 121:245–257

Producing fertile somatic hybrids

Edward C. Cocking

Plant Genetic Manipulation Group, Department of Botany, University of Nottingham, University Park, Nottingham NG7 2RD, UK

Abstract. For introgression of alien genes into a cultivated genome it is necessary for the somatic hybrids to possess some fertility. The use of fluorescence-activated cell sorting for the production of large numbers of somatic hybrid plants should enable a realistic analysis in this respect. Various factors, including the degree of aneuploidy and homology between chromosomes, as well as the ploidy levels of the parental species, affect fertility in somatic hybrids. Irradiation of one of the parental species to inactivate the nucleus is known to facilitate the production of cybrid plants: this can result in the production of fertile interspecies cybrids possessing novel cytoplasmic features. Irradiation-induced chromosome breakage can also facilitate limited gene transfer and can result in the production of transgenic fertile plants. Encouragingly, triploid plants resulting from the fusion of haploid gametic protoplasts with diploid protoplasts have been shown to possess high levels of fertility.

1988 Applications of plant cell and tissue culture. Wiley, Chichester (Ciba Foundation Symposium 137) p 75–89

Somatic cell hybridization is an extension of sexual hybridization. It overcomes sexual incompatibility barriers and creates a novel cytoplasmic mix as organelles of both parents come together in a common cytoplasmic milieu after protoplast fusion (Kumar & Cocking 1987). When methodological improvements are made in the culture and selection of hybrid cells arising from protoplast fusion, somatic hybridization can become part of general breeding programmes, provided the regenerated somatic hybrid plants have some fertility. Earlier studies on somatic hybridization by protoplast fusion emphasized the introduction of new genetic variability into plants by combining characters from distantly related species (Dudits & Praznovszky 1985). Currently, somatic hybridization for plant breeding improvements is increasingly utilizing species within the same genus or in closely related genera. Somatic hybrids provide unique opportunities to investigate cybridization, i.e. cytoplasmic transfer using protoplast fusion. This method permits the transfer and production of new combinations of cytoplasmic genes between species in one step, thereby circumventing a lengthy backcross programme. It also provides novel combinations in the case of both sexually compatible and sexually incompatible crosses (Fig. 1, Kumar & Cocking 1987). A distinct advantage of cybridization is that fertility is usually maintained in the cybrid plant, apart,

(A)

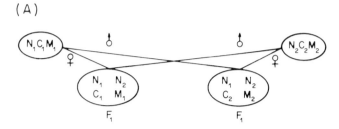

FIG. 1. Comparison of nuclear and cytoplasmic genomic combinations after reciprocal sexual hybridization and after somatic hybridization. (A) The F1 products of
reciprocal sexual hybridization in species with maternal inheritance.

of course, from the special case of the transference of cytoplasmic male
sterility by such somatic hybridization.

The synthesis of allotetraploids by the fusion of diploid somatic protoplasts
of different species is depicted diagramatically as $N_1 + N_2$ in Fig. 1. Familiar
examples of amphiploids obtained by sexual hybridization have been described by Cauderon (1977). These include the well known amphiploid
Raphanobrassica ($2n = 36$), first obtained by Karpechenko in 1927 from
spontaneous chromosome doubling of hybrids between *Raphanus sativus*
($2n = 18$) and *Brassica oleracea* ($2n = 18$). Recently, breeding trials have
shown the superiority of *Raphanobrassica* over rape for dry matter yield and
disease resistance, but fertility must still be improved. *Agropyron interme-
dium*, a perennial grass ($2n = 42$), the agronomically valuable characteristics
of which include resistance to wheat stem rust, leaf rust and yellow rust, can
be easily sexually crossed with wheats. *Agrotriticum* amphiploids have been
obtained by chromosome doubling of these F1s but their vigour and fertility
are very low. It is noteworthy that the primary *Triticale* directly resulting from
chromosome doubling of F1 sexual hybrids between hexaploid wheat and rye
generally had no significant agricultural value, mainly due to lack of fertility
and to shrivelling of the kernels. As discussed by Cauderon (1977), it was only
following observations on the efficiency of selection in the progenies of
crosses between different *Triticale* that breeding programmes were successfully undertaken. Indeed it has taken more than a century of breeding for this
'man-made' species to become an important crop plant.

One of the central challenges in most instances of somatic hybridization will
be the production of fertile somatic hybrid plants that can be incorporated
into a breeding programme. Significant advances are being made in this
respect and these are best illustrated by consideration of the improvement of
particular species by protoplast fusion.

Producing fertile potato somatic hybrids

There has been considerable interest in somatic hybridization as an additional

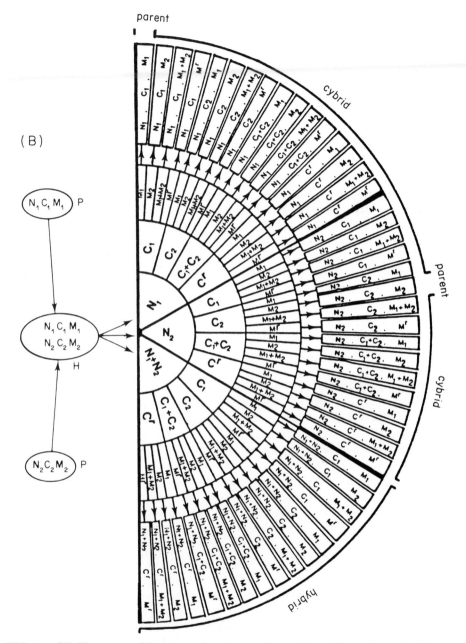

FIG. 1. (B) The range of fusion products possessing nuclear, chloroplast and mito-chondrial genomes which may arise through somatic hybridization after fusion of protoplasts of two species. P, protoplast; H, heterokaryon; F1, first filial generation; N, nucleus; C, chloroplasts; M, mitochondria; r, recombination; 1, one parent; 2, other parent.

method of improving potato cultivars, highlighted by the difficulty of applying conventional techniques to potato breeding because of their high level of sterility. Earlier attempts (Wenzel et al 1982) to produce clones with the optimal combination of qualitatively and quantitatively inherited traits by fusing protoplasts of carefully selected dihaploids from a conventional breeding programme with protoplasts of other dihaploids were unsuccessful. This was largely due to the lack of methods for the selection of fusion products. Riley (1979) has pointed out that somatic hybridization would gain a new dimension were this technique to be applied to haploids, especially monohaploids. For instance, if somatic hybrids could be produced by the fusion of potato monohaploids, there would be no need to double the chromosome number or to cross homozygous lines in order to determine combining ability. The fusion of protoplasts from two monohaploids would produce a diploid single-cross hybrid in one step. Furthermore, the breeder would not be dependent on the flowering and fertility of the monohaploids.

There are two main reasons why these somatic methods have not as yet been implemented. Reproducible regeneration of plants from potato protoplasts from a range of cultivars remains a challenge, even though there has been some progress (Haberlach et al 1985). Direct isolation of fusion products using fluorescence-activated cell sorting (FACS) is now available, enabling the rapid isolation of large numbers of hybrid cells (Alexander et al 1985), and this procedure should overcome the problem of the lack of mutants for selection.

Tetraploid and hexaploid somatic hybrid plants have been produced from protoplast fusions between *Solanum brevidens*, a 2x non-tuber-bearing species, and 2x and 4x *S. tuberosum*, with the aim of transferring the potato leaf roll virus (PLRV) resistance of *S. brevidens* to Group Tuberosum materials, since *S. brevidens* normally cannot be crossed directly to *S. tuberosum*. The next step in the practical incorporation of PLRV resistance from *S. brevidens* into *S. tuberosum* is the sexual transfer of PLRV resistance from such somatic hybrids to conventional breeding lines. In this connection, as fully discussed by Ehlenfeldt and Helgeson (1987), pollen viability, meiotic configurations and the ability to cross these fusion hybrids is an important concern. Encouragingly, the fertility of fusion hybrids, particularly the hexaploids, was satisfactory when the hexaploids were backcrossed to *S. tuberosum*. PLRV resistance from *S. brevidens*, and late blight resistance from *S. tuberosum*, can both be expressed in individual hybrids (Helgeson et al 1986). The chances, therefore, of incorporation of these genes into useful breeding lines are good. Ehlenfeldt and Helgeson (1987) suggested that the ratio of maternal to paternal chromosomes was important in allowing the endosperm of the offspring of the backcrossed somatic hybrid to develop. This factor is known to be important in the sexual crossing of *Solanum* species (Johnston & Hanneman 1982), and may be important in determining the fertility of somatic hybrids. As discussed below in connection with the production of fertile tobacco and petunia somatic hybrids, fusions between somatic protoplasts and pollen tetrad protoplasts to produce gametosomatic triploids could be advantageous. Only future research will determine whether hexaploids produced from triploids between these two potato species are advantageous for fertility.

Producing fertile tomato somatic hybrids

Species of the genus *Lycopersicon* can be divided into two major groups on the basis of their sexual crossing relationships: one group can be hybridized with the cultivated tomato with relative ease (the *'esculentum'* complex), the other group is separated from the cultivated species by severe sexual barriers (the *'peruvianum'* complex). The species of the *'peruvianum'* complex represent a vast store of potentially useful genetic variation, which tomato breeders are frequently deterred from utilizing because of the daunting sexual barriers that have to be overcome. Even when sexual hybrids have been produced, they have frequently lacked fertility.

Somatic hybrid plants have been obtained between *L. peruvianum* and *L. pennellii* but these hybrids were sterile (Adams & Quiros 1985). Our studies in the Plant Genetic Manipulation Group at Nottingham have centred on investigating the consequences of fusion of protoplasts from leaves of *L. esculentum* with protoplasts from cell suspension cultures of *L. peruvianum*. Our assessment of work so far on somatic hybridization is that, in most instances, insufficient numbers of heterokaryons have been selected and regenerated into whole plants for evaluation of hybrid fertility. We therefore undertook a large number of experiments (twenty-one), in each of which there was a satisfactory fusion frequency. From only three of these were somatic hybrids produced. As previously mentioned, the use of FACS may resolve this problem of numbers.

The selected somatic hybrid plants *L. esculentum* (+) *L. peruvianum* ($2n = 6x = 72$) were hexaploid, fertile and set seed after self pollination (Kinsara et al 1986). This result is encouraging for future use of protoplast fusion for the introgression of desired genes from the *'peruvianum'* complex into the cultivated *'esculentum'* complex. This is an example where protoplast fusion has resulted in the production of fertile somatic hybrid plants, when sexual crosses (albeit produced with difficulty) in this important crop usually do not produce fertile plants. Moreover, since the variety of cultivated tomato utilized is a non-regenerating variety from protoplasts (Morgan & Cocking 1982), these procedures (preferably coupled with FACS selection) will probably be applicable to a wide range of other non-regenerating cultivated tomato varieties and regenerating wild-type *Lycopersicon* species. It may also be useful to explore the breeding behaviour of any gametosomatic hybrids (*L. esculentum* (x) *L. peruvianum*), or of pollen tetrad protoplasts of wild species other than *L. peruvianum*, when backcrossed with *L. esculentum*. This would avoid high ploidy levels in hybrids and be an asset in current breeding programmes to improve the cultivated tomato.

Producing fertile petunia and tobacco somatic hybrids

Tobacco is the most widely grown commercial non-food plant in the world, and it is understandable that it has attracted attention with respect to the application of protoplast fusion methodology. In addition, many species of tobacco are particularly amenable to the regeneration of plants from protoplasts. Significant advances in the development of cytoplasmic male sterility via interspecific hybridization have been made by Kubo (1985) using protoplast fusion methodology. Until recently, cytoplasmic male sterile lines in tobacco were produced by the backcross method but, as pointed out by Kubo (1985), the time required for cytoplasmic substitution by the backcross method is comparatively long. This is a major disadvantage in tobacco hybrid breeding, for which the rapid production of both new varieties and male sterile lines is a prerequisite. In tobacco, only about one year was required for the production of male sterile lines by protoplast fusion, whereas the backcross method required at least 3–4 years (Kubo 1985).

Hamill et al (1985) have carried out a detailed investigation of the fertility in somatic hybrids of *Nicotiana rustica* and *N. tabacum*, and their progeny over two sexual generations. Somatic hybrid plants produced between *N. rustica* and *N. tabacum* by heterokaryon isolation and culture, and also by mutant complementation, were examined for their ability to set seed. Much variation in fertility was observed in subsequent generations; by recurrent selection of the most fertile over two generations, it was possible to increase the level of self-fertility in some of the progeny. All the *N. rustica* (+) *N. tabacum* somatic hybrids analysed were aneuploid, showing that aneuploidy does not necessarily result in sterility. It is possible that such a strategy of recurrent selection for increased self-fertility in somatic hybrids, over several generations, may allow the detection of individuals most suitable for backcrossing with parental species in plant breeding programmes. Anther culture of somatic hybrids might also be particularly useful in this respect. Again the use of FACS to produce much larger numbers of somatic hybrids would clearly be advantageous. Although somatic hybrids with self-fertility may hasten the introgression of characters from a somatic hybrid into a crop species, self-fertility of somatic hybrids is not an absolute prerequisite for the use of a plant in a breeding programme. Backcrossing of *N. tabacum* and *N. mesophilia* somatic hybrids, with limited fertility, to the *N. tabacum* parent has been carried out with the aim of transferring disease resistance into tobacco (Evans et al 1981). Backcrossing of somatic hybrids between *N.*

rustica and *N. tabacum* with a high nicotine content (with little or no self-fertility) to a commercial cultivar of *N. tabacum* produced, after several generations, acceptable levels of seed upon selfing (Pandeya et al 1986). These offspring possessed significantly increased levels of nicotine compared with parental types, and black root rot resistance inherited from the *N. rustica* parent.

The possibility of the transfer of only part of the genome in somatic hybridization in tobacco, and thereby the facilitation of the maintenance of fertility, has been examined extensively. For instance, Bates et al (1987) electrically fused mesophyll protoplasts of a kanamycin-resistant, nopaline-positive *N. plumbaginifolia* inactivated by γ-irradiation with unirradiated mesophyll protoplasts of *N. tabacum*. Somatic hybrids were selected by their capacity to grow in the presence of kanamycin. The irradiation caused extensive elimination of *N. plumbaginifolia* genetic material, resulting in the formation of highly asymmetric somatic hybrids. These asymmetric hybrids were female fertile to varying degrees, and after backcrossing these hybrids with *N. tabacum*, inheritance to kanamycin-resistance was biased toward sensitive progeny. Interestingly, immunological evidence has been obtained for the transfer of the barley nitrate reductase structural gene to *N. tabacum* by protoplast fusion (Somers et al 1986). The fusion of *N. tabacum* somatic mesophyll protoplasts (in this instance the nitrate reductase-deficient mutant) with *N. glutinosa* pollen tetrad (gametic) protoplasts has led to the production of allotriploids (*N. glutinosa* (x) *N. tabacum*) plants, which possess good self- and cross-fertility (Pirrie & Power 1986). The ability to produce such inter-specific triploids possessing the haploid genome of the alien species may facilitate the transfer of only part of the genome (perhaps one or a few chromosomes) from an alien species, e.g. *N. glutinosa*, into the cultivated species, e.g. *N. tabacum* (Pental & Cocking 1985).

Analysis of an aneuploid somatic hybrid *Petunia parodii* ($2n = 14$) and *P. parviflora* ($2n = 18$) with 31 chromosomes has shown that the hybrid behaves as an allotetraploid. This hybrid between these two sexually incompatible species was sterile. A detailed chromosomal analysis showed that there had been an interchange of segments between one of the large satellited *P. parodii* chromosomes and a *P. parviflora* chromosome of medium size, followed by loss of one of the interchange products (White & Rees 1985). Recently, somatic hybrid plants have been produced by fusion between *N. tabacum* protoplasts and *P. inflata* protoplasts, resulting in the production of the amphiploid hybrid (*N. tabacum* (+) *P. inflata*). This 'Nicotunia' hybrid ($2n = 6x = 62$) has some fertility in backcrosses with *N. tabacum* (Kim et al 1988).

The outlook for the production of fertile somatic hybrids in other genera

Somatic hybrids between *Brassica oleracea* and *B. campestris* have been produced by the use of iodoacetamide inactivation (Terada et al 1987a). Some of the hybrids flowered and set seeds, even though chromosome analysis had revealed a considerable variation in the chromosome number of the somatic hybrids, showing that multiple fusions had occurred followed by chromosome loss during culture. Now that it is possible to regenerate plants efficiently and reproducibly from rice protoplasts through somatic embryogenesis (Abdullah et al 1986), it is likely that somatic hybrids (balanced, asymmetric and gameto) will be produced between *Oryza sativa* and wild rice species, and also male sterile lines by protoplast fusions. Encouragingly, plantlet regeneration from somatic hybrids of rice and barnyard grass has been reported (Terada et al 1987b).

The steady increase in the capability of plant regeneration from protoplasts of a wide range of species is providing the required foundation for the production of somatic hybrid plants by protoplast fusion. If this is coupled with efficient methods for the selection of large numbers of hybrid cells, especially the use of fluorescence-activated cell sorting, then the production of fertile somatic and gametosomatic hybrids with applications in many aspects of plant breeding will be assured.

References

Abdullah R, Cocking EC, Thompson JA 1986 Efficient plant regeneration from rice protoplasts through somatic embryogenesis. Bio/Technol 4:1087–1090

Adams TL, Quiros CF 1985 Somatic hybridization between *Lycopersicon peruvianum* and *L. pennellii*: regenerating ability and antibiotic resistance as selection systems. Plant Sci 40:209–219

Alexander RG, Cocking EC, Jackson PJ, Jett JH 1985 The characterization and isolation of plant heterokaryons by flow cytometry. Protoplasma 128:52–58

Bates GW, Hasenkampf CA, Contolini CL, Piastuch WC 1987 Asymmetric hybridization in *Nicotiana* by fusion of irradiated protoplasts. Theor Appl Genet 74:718–727

Cauderon Y 1977 Alloploidy. In: Sanchez-Monge E, Garcia-Olmedo F (eds) Interspecific hybridization in plant breeding. Proc eighth Congr Eucarpia. Madrid, Spain

Dudits D, Praznovszky T 1985 Intergeneric gene transfer by protoplast fusion and uptake of isolated chromosomes. In: Zaitlin M et al (eds) Biotechnology in plant science, relevance to agriculture in the eighties. Academic Press, London p115–127

Ehlenfeldt MK, Helgeson JP 1987 Fertility of somatic hybrids from protoplast fusions of *Solanum brevidens* and *S. tuberosum*. Theor Appl Genet 73:395–402

Evans DA, Flick CE, Jensen RA 1981 Disease resistance: Incorporation into sexually incompatible somatic hybrids of the genus *Nicotiana*. Science (Wash DC) 213:907–909

Haberlach GT, Cohen BA, Reichert NA, Baer MA, Towill LE, Helgeson JP 1985 Isolation, culture and regeneration of protoplasts from potato and several related *Solanum* species. Plant Sci Lett 39:67–74

Hamill JD, Pental D, Cocking EC 1985 Analysis of fertility in somatic hybrids of *Nicotiana rustica* and *N. tabacum* and progeny over two sexual generations. Theor Appl Genet 71:486–490

Helgeson JP, Hunt GJ, Haberlach GT, Austin S 1986 Somatic hybrids between

Solanum brevidens and *Solanum tuberosum*: Expression of a late blight resistance gene and potato leaf roll resistance. Plant Cell Rep 3:212–214

Johnston SA, Hanneman Jr RE 1982 Manipulations of endosperm balance number overcome crossing barriers between diploid *Solanum* species. Science (Wash DC) 217:446–448

Karpechenko GD 1927 Polyploid hybrids of Raphanus sativus L. × Brassica oleracea L. Bulletin of Applied Botany of Genetics of Plant Breeding (Number 3) Vol XVII p 398–408

Kim JC, Cocking EC, Power JB 1988 Somatic hybrids of *Nicotiana tabacum* and *Petunia inflata* by protoplast electrofusion. Plant Sci, submitted

Kinsara A, Patnaik SN, Cocking EC, Power JB 1986 Somatic hybrid plants of *Lycopersicon esculentum* Mill. and *Lycopersicon peruvianum* Mill. J Plant Physiol 125:225–234

Kubo T 1985 Bulletin. IWATA Tobacco Exp Stn No. 17:130–133

Kumar A, Cocking EC 1987 Protoplast fusion: a novel approach to organelle genetics in higher plants. Am J Bot 74:1289–1303

Morgan A, Cocking EC 1982 Plant regeneration from protoplasts of *Lycopersicon esculentum* Mill. Z Pflanzenzuecht 106:87–104

Pandeya RS, Douglas GC, Keller WA, Setterfield G, Patrick ZA 1986 Somatic hybridization between *Nicotiana rustica* and *N. tabacum*: development of tobacco breeding strains with disease resistance and elevated nicotine content. Z Pflanzenzuecht 96:346–352

Pental D, Cocking EC 1985 Some theoretical and practical possibilities of plant genetic manipulation using protoplasts. Hereditas (Suppl) 3:83–92

Pirrie A, Power JB 1986 The production of fertile, triploid somatic hybrid plants (*Nicotiana glutinosa* (n) + *N. tabacum* (2n) via gametic-somatic protoplast fusion. Theor Appl Genet 72:48–52

Riley R 1979 Breeding for symbiotic nitrogen fixation. In: Sneep J, Hendriksen AJT, (eds) Plant breeding perspectives. UNIPUB, New York

Somers DA, Narayanan KR, Kleinhofs A, Cooper-Bland S, Cocking EC 1986 Immunological evidence for transfer of the barley nitrate reductase structural gene to *Nicotiana tabacum* by protoplast fusion. Mol & Gen Genet 204:296–301

Terada R, Yamashita Y, Nishibayashi S, Shimamoto K 1987a Somatic hybrids between *Brassica oleracea* and *B. campestris*: selection by the use of iodoacetamide inactivation and regeneration ability. Theor Appl Genet 73:379–384

Terada R, Kyozuka J, Nishibayashi S, Shimamoto K 1987b Plantlet regeneration from somatic hybrids of rice (*Oryza sativa* L.) and barnyard grass (*Echinochloa oryzicola* Vasing). Mol & Gen Genet 210:39–44

Wenzel G, Meyer C, Przewozny T, Uhrig H, Schieder O 1982 In: Earle ED (ed) Variability in plants regenerated from tissue culture. Praeger, New York, p 290–302

White J, Rees H 1985 The chromosome cytology of a somatic hybrid petunia. Heredity 55:53–59

DISCUSSION

Harada: Professor Cocking, could you please define the term introgression? What is the difference between introduction and introgression, as you use it?

Cocking: I was using the term introgression in a rather wide sense of giving either chromosome addition lines in progeny or actual crossover of chromosomes to give incorporate groups of genes from an alien species.

Galun: The actual chromosomal status of these somatic hybrid plants is pentaploid but you talk about triploid. How would you define these, as triploid or pentaploid?

Cocking: This is difficult terminology because of the ploidy status of *Nicotiana tabacum*, which is allotetraploid, and *N. glutinosa*, which is diploid. I would say it was an inter-species triploid, but you might disagree.

Scowcroft: I think the functional behaviour is quite different. With your material you are basically dealing with a complete, intact genome set for tobacco, and the pairing relationships in meiosis will be in order; the only thing that will be out of order is the haploid set from *N. glutinosa*. In a true triploid, functionally you have three sets of chromosomes and pairing relationships during meiosis will be perturbed. In your case, I don't think that you can use what happens in the production of a normal triploid as an analogy. The *N. tabacum* tetraploid genome is intact and you have added a haploid genome set from *N. glutinosa*.

Riley: Cytogeneticists have a lot of experience with plants of the structure that that you describe, Ted, and I think generally they would call them pentaploids. For example, the hybrid between *Triticum aestivum* and *T. durum* would have essentially the structure that Bill (Scowcroft) has been describing, namely four intact genomes and one single genome, i.e. one genome where every chromosome is in the monosomic state, but two genomes where every chromosome is in the disomic state. The fertility is greatly buffered by the fact that the *N. tabacum* chromosomes segregate perfectly normally at meiosis, and the only odd balls are the monosomic chromosomes of the *N. glutinosum* set. The triploids may be of various kinds—in an allotriploid two genomes are the same while one is different; an autotriploid may have three genomes all the same. In those instances, meiosis is greatly distorted and segregation into the gametes is extremely irregular but this is not the case in the plant you describe.

Hall: Do you intend to do any studies on the chloroplast proteins, since one of the most intriguing situations is the contribution of the nucleus to the function of the chloroplast and the integration of nuclear gene products. It might be quite interesting to see whether your chloroplasts integrate different proteins coded by your different genomes.

Cocking: We have had a preliminary look at the chloroplast status of this plant 'material'. As one might expect from its origin, this gametosomatic hybrid material had only the chloroplast genome of *N. tabacum*.

Hall: The question goes beyond that: within the chloroplast there are nuclear encoded proteins, and one wonders whether there is competition between the proteins that are encoded in the various nuclear contributions to your plant material.

Cocking: We haven't looked at that level in this material.

Potrykus: I like this general idea of using non-dividing tetrad protoplasts. My own experience is that they do not divide. It provides a generally applicable partially selective system for recovering hybrids. The question is, how certain is it that these 'gametosomatic hybrids' are really hybrids? It would be good to get additional evidence, because cells with sixty chromosomes can be obtained from cultures of *N. tabacum* alone. We have seen that you can get reversion of nitratereductase-deficient *N. tabacum*. So the possible alternative hypothesis would be that what you have recovered are not 'gametosomatic hybrids' but a reversion of *N. tabacum* cell culture protoplasts to a sixty chromosome variant. Could you exclude this possibility?

Cocking: I would exclude the possibility because of the results of the isoenzyme analyses that have been carried out. Isoenzyme analysis of esterase, peroxidase and Fraction 1 protein polypeptides all support the hypothesis that this is hybrid material. In this respect it has the same status as all other authenticated somatic hybrid material.

Potrykus: So we have not seen all the evidence you have?

Cocking: The detailed evidence for the hybridity of this material has been published (Pirrie & Power 1986).

Potrykus: I think it's a very nice general system to apply because I know how easy it is to get pollen tetrad protoplasts and you have a half-selected system.

Cocking: It's clear that the pollen tetrad nucleus probably is activated to division by being in the mixed cytoplasm of itself and the fusion partner. That in itself is fascinating, and might give a clue in the future as to how one could activate pollen to undergo repeated mitotic division. If we had a better understanding of the nuclear interactions leading to the triggering of division in the pollen tetrad nucleus, we might be able to programme the development of pollen tetrads in culture. The division of the pollen tetrad nucleus may be triggered by a cytoplasmic factor from the somatic system, or it may occur even if the somatic nucleus was irradiated. So I think there are a number of basic cell biological possibilities there that may enhance studies in related fields as well as providing novel material for hybridization.

Potrykus: Because of all these reasons, it is especially important to have absolutely tight evidence.

Cocking: The evidence is absolutely tight!

Galun: This system is potentially very useful for certain applications. In many fruit trees the breeders are very interested in seedlessness, such as apples and citrus fruits. Your system gives us the potential to produce real triploids by just fusing diploid protoplasts with tetrad protoplasts. The only limitation, if this fusion is achievable, is that in addition to seedlessness there should also be a parthenocarpy gene involved. This gene is already present in many citrus trees, so there is a good chance that we could achieve in one stroke the production of seedless triploids.

Scowcroft: I don't think you are right. I would agree with you, if you were talking about somatic hybridization between two haploid gametophytes to give a true triploid, but in this particular case you are fusing a haploid genome with a diploid or tetraploid genome—the pairing relationships in one partner are intact during subsequent meiosis, and so you will get seeds out of it. It's only when those pairing relationships during meiosis are destroyed that seedless fruits will be produced.

Riley: Japan is the country in which the seedless watermelon was invented by Hitoshi Kihara and his colleagues. The seedless watermelon depends upon fertilization of a tetraploid watermelon by a diploid watermelon. It puts together three sets of chromosomes, the meiotic pairing between which is so confused, and the segregation so distorted, that little or no seed is produced.

Galun: We are talking about homologous fusion in citrus and apples, in which case I suggest that the protoplasts come from the same species—it's a fusion between a tetrad protoplast and a normal protoplast from the same species.

Potrykus: But why would you want to do that fusion when it is quite easy to cross plants of the same species?

Galun: It's not easy to produce triploids, this is one of the best ways to produce triploids.

Riley: Chairman, may I say how much I enjoyed listening to Teds' paper. It put across the position of inter-species crossing by protoplast fusion particularly well. I was especially interested to hear him say that those who work in this field are more and more looking to hybridizations within genera rather than between genera. I wonder what he feels about these wider hybridizations. What has impressed me from the work of some of my colleagues is, even in very wide sexual hybrids, the way in which chromosome elimination takes place. For example, in maize x wheat hybrids there is rapid elimination of the maize genome early in embryo development. Is there any example of that in somatic hybridization?

Galun: There is recent work by Dudits (1987), where he had such a case of elimination of one of the parental chromosomes and production of a fertile plant.

Riley: Are you able to tell us what genus?

Galun: One of them is obviously carrot, the other one is tobacco.

Potrykus: Was there good evidence that these were hybrids at the start of the experiment? There are two possibilities: that there was a hybrid and chromosomes were lost, or that there was not a hybrid in the first place.

Galun: It looks OK to me.

Zenk: I think this *Nicotiana* would be a nice model in which to study the chemistry. Both parent species make alkaloids, I think you should ask a chemist to analyse the hybrid by gas chromatography and see how the pattern of the parents compares with that from the somatic hybrids. It could be that there is evolution of some hybrid alkaloids in these plants.

Cocking: We have begun a little of that type of work in other *Solanum* species. This material is freely available; if any biochemists would like to do this analysis, I would be only too pleased.

Scowcroft: When you are trying to establish hybridity in some of these protoplast fusion type experiments, for example by isozyme analysis, do you ever find cases where you find proof of hybridity with one pair of isozymes but not with another pair? I am thinking about one of the things that happens in some of the interspecific hybrids produced sexually in wheat: there is a phenomenon called gene silencing, the presence of one genome will silence the expression of a gene in another genome. Has this ever been observed in plants derived by somatic hybridization?

Cocking: Not that I have seen published. There is the danger in this field that if people get results that they don't fully understand, they tend not to include such results in their publications.

Potrykus: This sounds like a nasty comment, if people would take the verification of hybrids more seriously then we might already know the answers to this question. There are many 'hybrids' in the literature, where the evidence that they are true hybrids is very poor.

Yamada: In our laboratory we have a hybrid shoot culture between *Duboisia hopwoodii* and *N. tabacum* (Endo et al 1987). We analysed the karyotype of the hybrid cell line and regenerated shoots: it is very easy to distinguish from which plant the chromosomes came by the size of the chromosomes. In *D. hopwoodii*, 2n equals 60, while in tobacco 2n equals 48. We found more than 200 chromosomes after fusion in metaphase figures of the hybrid cell line. Both *Duboisia* and *Nicotiana* chromosomes were clearly present in these division figures. Professor Zenk asked about secondary metabolite production in the hybrid plants. Tobacco produces nicotine; *D. hopwoodii* produces both nicotine and tropane alkaloids. So far, our hybrid shoots have not formed roots, so we have yet to analyse alkaloids of the hybrid (T. Endo, T. Komiya, Mino, Y. Yamada, Nakanishi and Y. Fujita, unpublished work).

Zenk: You don't need a root for production of nicotine and scopolamine.

Yamada: I think that root formation is necessary, because these substances are made in the root and then transported to the leaves.

Zenk: I think both tissues can make it independently.

Potrykus: There is a neutral, generally applicable technique to see whether a presumptive hybrid is really a hybrid: it is possible to isolate species-specific DNA sequences which do not cross-hybridize to both fusion parents and these can be used to see whether DNA from both fusion partners is present in the presumptive hybrid (Saul & Potrykus 1984).

Cocking: I agree that the more biochemical analysis one can do on material, the better. It's interesting that in the development of this field, the first thing people were worried about was whether the system was chimaeric, this particularly applied to callus material. I think that even your refined molecular methods will not necessarily prove that you have a hybrid at the callus level because it could be chimaeric. Even with more sophisticated analysis, you have to be very careful in relation to the status of the material. Interestingly, I know of no situation where a chimaeric plant has been produced as a result of attempted somatic hybridization. Dr Don Cheney told me recently that he had produced a chimaeric seaweed by attempting protoplast fusion, and producing not a somatic hybrid but a chimaera. I agree wholeheartedly with the tenor of your comments that more analytical proof is required, whether it be at the DNA level, at the isoenzyme level or at the secondary plant product level. I think that in somatic hybrids there is a great opportunity, as our Chairman is studying, to try to get the best of both worlds in relation to secondary product synthesis and possibly isolate some hybrid type molecules at the secondary product level.

Yamada: It's very difficult because we usually make hybrids by fusing cultured cells which are very heterogeneous in their potential for secondary metabolite production. So even after we have produced a hybrid, we don't know exactly what kind of hybrid it is from the metabolic pathway.   I also disagree with Professor Zenk: I think that it's difficult to induce root formation and rather easier to induce shoot formation. Generally people think that root formation is much easier to achieve than shoot formation, but in the case of the hybrid between *Duboisia* and *N. tabacum*, we have the reverse condition. We have been culturing hybrid shoots for two years. We have tried root induction by *Agrobacterium rhizogenes*-mediated transformation, as well as by other methods, but so far roots have not been induced. We have not found any alkaloids in the hybrid, unfortunately.

Takebe: Have you ever looked, Professor Cocking, at your gametosomatic hybrids between *N. glutinosa* and *N. tabacum*, to see whether they respond to infection by tobacco mosaic virus with the hypersensitive reaction?

Cocking: I feel very guilty to admit that I have not, despite my previous extensive work on tobacco mosaic virus.

Yamada: I think my question to Professor Cocking is more general and perhaps I should ask it of everyone here. A hybrid made by cell fusion can be propagated by meristem culture very easily, even if it is not fertile. Of course from the scientific point of view it is very important and very useful to have a fertile plant, but from the practical viewpoint, I think vegetative propagation of a hybrid is not difficult. Do you think that meristem propagation can lead to the appearance of variants? Do you think that it is necessary to produce hybrids that can be propagated by traditional breeding methods?

Cocking: I'll respond in general terms Professor Yamada by recounting our experiences. We generated a somatic hybrid between *Petunia parodii* and *P. parviflora*, which had what we call in England a 'hanging basket' habit. We thought that we could sell this if we could propagate it readily, which we can do by vegetative propagation. It is stable, you don't have to use meristem culture.

We tried to sell it but the flower, although attractive, was very small. *P. parodii*, the white flowering Petunia, is not as attractive as many of the *P. hybrida*. We have now produced the hybrid between *P. parviflora* and *P. hybrida*. This has a much showier flower and is suitable for hanging baskets for Japanese houses, certainly for Swiss houses! We are going to link up with a commercial firm to sell vegetatively propagated material. Admittedly the housewife may vegetatively propagate it herself and therefore there's limited protection, but at least it's a novelty and that's what I think will count. But we haven't encountered major chromosomal changes, largely because we have not put the material back into tissue culture, but only used soil propagation.

Barz: Professor Cocking, could you comment on the applicability of your fusion system to transfer resistance against a microbial pathogen from a cultivar into a hybrid? One could visualize a situation where this could be a good method especially if you are dealing with a non host-specific toxin. Has that approach been used?

Cocking: Yes, a good example of this is the cross between *N. rustica* and *N. tabacum*. The somatic hybrid has been extensively investigated by Canadian colleagues, (Dr R.S. Pandeya, Dr G.C. Douglas, Dr W.A. Keller, Dr G. Setterfield, Dr Z.A. Patrick) They have backcrossed the somatic hybrid to *N. tabacum* and have obtained resistance in their tobacco to the fungus that causes black shank disease. This is a typical example. The aim of the plant breeders in many of these crosses is to enhance disease resistance in the cultivated species. This is certainly the case in programmes on somatic hybridization in rice. These are concentrating on resistance to the insect vector, if the insect is unable to insert its proboscis into the plant, it can't transfer the virus. So the programmes being developed are in connection with inter-species hybridization in the oryza genus, where there have been major problems in sexual crossing, and where it is now possible to regenerate plants from rice protoplasts. There's a whole spectrum of such objectives in these broader hybridization programmes where one knows that there is a germplasm pool of such disease resistance available.

References

Dudits D 1987 Transfer of resistance traits from carrot into tobacco by asymmetric somatic hybridization: regeneration of fertile plants. Proc Natl Acad Sci USA 84:8434–8439

Endo T, Komiya T, Masumitsu Y, Morikawa H, Yamada Y 1987 An intergeneric hybrid cell line of *Duboisia hopwoodii* and *Nicotiana tabacum* by protoplast fusion. J Plant Physiol 129:453–459

Pirrie A, Power JB 1986 The production of fertile, triploid somatic hybrid plants (*Nicotiana glutinosia* (n) + *N.tabacum* (2n)) via gametic:somatic protoplast fusion. Theor Appl Genet 72:48–52

Saul MW, Potrykus I 1984 Species-specific repetitive DNA used to identify interspecific somatic hybrids. Plant Cell Rep 3:65–67

Application of microinjection to a high frequency and synchronous somatic embryogenesis system in carrot suspension cultures

Atsushi Komamine and Koji Nomura

Biological Institute, Faculty of Science, Tohoku University, Sendai, 980 Japan

Abstract. Microinjection is a useful technique both for investigating the mechanisms of somatic embryogenesis (which can be regarded as a model of totipotency) and for obtaining cloned transgenic plants at high frequency. In this paper we report 1) the establishment of a high frequency somatic embryogenesis system from single cells, 2) the establishment of microinjection techniques for single plant cells with cell walls, 3) analysis by microinjection techniques of the mechanisms of totipotency expression, and 4) our attempts to produce cloned transgenic plants using microinjection techniques.

1988 Applications of plant cell and tissue culture. Wiley, Chichester (Ciba Foundation Symposium 137) p 90–96

Somatic embryogenesis is an ideal system for investigating mechanisms of gene expression, as well as for obtaining cloned plants, because a whole plant can be differentiated from a single cell.

The aim of the work described in this paper was to analyse the mechanisms of somatic embryogenesis in a high frequency and synchronous embryogenesis system of carrot suspension cultures using microinjection techniques. We also attempted to obtain transgenic cloned plants in the same system by the introduction of foreign DNA using microinjection.

A high frequency and synchronous somatic embryogenesis system was established by selection of highly totipotent, small, round, cytoplasm-rich single cells. This was done by sieving with nylon screens (pore size 10–16 μm), followed by density gradient centrifugation in Percoll solutions. Selected single cells were cultured in a medium containing 2,4-dichlorophenoxyacetic acid (2,4-D), then transferred to one containing zeatin, a cytokinin, but lacking auxin. Single cells differentiated into embryos and subsequently into plantlets at a frequency of 85–90% in this system (Nomura & Komamine 1985).

A microinjection method was established for intact single cells with cell walls using the high frequency embryogenesis system described above. A solution of a flourescent dye, Lucifer Yellow CH, was microinjected into those single cells, using an inverted microscope and a hydraulic micromanipulator. In order to hold cells with cell walls and to overcome their turgor pressure, certain modifications to conventional microinjection methods for protoplasts were necessary. The microinjected cells could divide at a frequency of about 70% and about 50% could differentiate (Nomura & Komamine 1986).

There are two types of single cells in the carrot suspension cultures used in the present study: Type 1 are small, round, highly totipotent single cells which give rise to embryogenic cell clusters in medium containing auxin and then embryos when transferred to a medium lacking auxin. Type 2 are elongated single cells, which do not divide and differentiate into embryos under any conditions tested so far. The latter type of cells seem to have lost totipotency. When Type 1 single cells were cultured in the medium lacking auxin, they became elongated, showed a similar shape to Type 2 cells and could not differentiate into embryos even in the culture medium containing auxin.

Smith et al (1985) isolated a monoclonal antibody, 21D7, against a protein specific for embryogenesis. We added 21D7 to our system and found that the protein which cross reacted with 21D7 (called 21D7 protein) could be detected in Type 1 cells and during the process of embryogenesis from Type 1 cells. However, 21D7 protein could not be detected in Type 2 cells.

When the antibody 21D7 was microinjected into Type 1 cells, these cells could no longer divide or differentiate and their morphology became similar to that of Type 2 cells. This indicates that 21D7 protein can be regarded as a marker protein for totipotency and may play an important role in the early stage of embryogenesis.

Lucifer Yellow CH was microinjected into embryogenic cell clusters and pattern formation was investigated by following distribution of the fluorescent dye. The results obtained indicated that the fate of each cell in embryogenic cell clusters was predetermined. In other words, determination of embryogenesis occurs during the formation of the embryogenic cell clusters from single cells.

If single cells can be transformed by the introduction of foreign DNA using microinjection, then this embryogenesis system can be used to allow the transformed cells to differentiate at a high frequency, thereby producing large numbers of transgenic clone plants. A DNA fragment of the Ti plasmid from *Agrobacterium tumefaciens*, on which the *ocs* gene, which encodes lysopine dehydrogenase required for the production of octopine, was located, was microinjected into Type 1 single cells. After regeneration to plantlets via embryogenesis, it was confirmed by Southern hybridization using a DNA chemiprobe kit that the introduced foreign DNA was replicated. Detection of octopine biosynthetic activity using labelled precursor, arginine, and protein analysis by gel electrophoresis showed that the introduced gene was expressed in the plantlets. The frequency of expression of the foreign DNA in regenerated plants from single cells was about 5%.

These results indicate that microinjection is a useful technique for analysis of the mechanisms of somatic embryogenesis as well as for obtaining cloned transgenic plants at high frequency.

References

Nomura K, Komamine A 1985 Identification and isolation of single cells that produce somatic embryos at a high frequency in a carrot suspension culture. Plant Physiol (Bethesda) 79:988–991
Nomura K, Komamine A 1986 Embryogenesis from microinjected single cells in a carrot cell suspension culture. Plant Sci 44:53–58
Smith JA, Choi JH, Kraus MR, Karu AE, Sung ZR 1985 Monoclonal antibodies that recognize proteins unique to somatic embryos of Daucus carota. In: Terzi M et al (eds) Somatic embryogenesis. IPRA, Rome, p 86–94

DISCUSSION

Cocking: Have you been able to extend this microinjection approach to any species other than the present carrot system? One of the difficulties may be to imitate this approach at the somatic suspension culture level with other species. In that connection, might it not be better to use developing microspores, from which plants could be regenerated, as a starting system, because my impression is that the cultured isolated pollen system could be much more readily applied than the system that you have described, and might offer other advantages as well?

Komamine: We have not yet tried other systems. I agree that microinjection of microspores would be a fruitful approach to breeding. We must establish a high frequency embryogenesis system, whether with microspores or somatic cells of other crops. We are now trying to establish a high frequency embryogenesis system in rice but we have not yet succeeded.

Cocking: It might be easier to establish a rice microspore system or a wheat microspore system or the barley microspore system which we have already heard about, than to do this microinjection at the level of normal somatic cells in suspension culture. That's the impression I get from talking to people and from the literature.

Komamine: Yes, but in practice it is rather difficult to get microspores which are at the stage suitable for the induction of embryogenesis.

Potrykus: This system is beautiful but it is unfortunately limited to carrot at this moment. We also felt that embryogenic cereal microspores might be a valid system in which to test whether one could transfer genes by microinjection. This has been tried by Neuhaus and Spangenberg, first with the very embryogenic microspores in *Brassica*, where it has been shown to work. The technique is relatively simple. Since one has not such a high frequency of regeneration from single cells as in carrot, one does not use single cells but early pro-embryos at the 12–16 cell stage. Microinjection of these produces transgenic chimaeras (Neuhaus et al 1987). We have applied this technique to rice, wheat, barley and maize, and we hope that this will produce transgenic chimaeras again. We would, however, like to wait until we have the sexual offspring from these plants and definite proof before we claim too strongly.

Withers: In situations where there is no embryogenic cell or microspore system available, might it be possible to use an early stage zygotic embyro or even a shoot tip? There would be a risk of producing a chimera, of course, but with suitable markers that need not be a problem.

Komamine: That is a possibility.

Hall: I would like to say Professor Komamine, what impressive and convincing work you have presented. Could you just clarify, did the injection of the 21D7 monoclonal inhibit the embryogenesis?

Komamine: Yes.

Hall: So the cells that were microinjected with the antibody would go to proliferation for rooting but they would not go to embryogenesis?

Komamine: If we microinject 21D7 monoclonal antibody, this will cross react with the antigen, 21D7 protein, which we think is important for embryogenesis, resulting in the inhibition of embryogenesis. That was our assumption when we tried the microinjection of 21D7 and we did get inhibition of embryogenesis by microinjection of this monoclonal antibody.

Hall: If you have indeed identified 21D7 as being specific against a protein involved in embryogenesis, then it should prevent the cells going through to embryogenesis but non-embryogenic division should still be possible. Did you observe non-embryogenic division, i.e. did the antibody specifically prevent embryogenesis?

Komamine: The process of embryogenesis includes cell division but some differentiation also occurs during in cell division.

Hall: Unfortunately, stopping cell division totally is rather easier than blocking a specific pathway.

Komamine: It is suggested only from the results presented here that the 21D7 antigen may play a role in the induction of cell division. We used immunocytological methods to investigate the distribution of 21D7 and found, for example, that the antigen is present in the meristem. However, we have a suspension culture of *Catharanthus roseus* which has been subcultured for more than sixteen years and has lost totipotency but in these cells we could hardly detect the 21D7 antigen (this work was done by Dr Jane Smith in my laboratory). So at this moment we can speculate that the 21D7 antigen may either play an important role in the expression of totipotency or be a marker protein for totipotency.

Potrykus: I was impressed by your experiments. It will, however, be important to determine whether or not your 21D7 protein is really an embryo-specific protein or a cell division-specific protein. Would it be possible to use the replicating carrot cell culture to determine whether this antigen is expressed in simple mitotically cycling cells? It is important to use the same species for both sets of investigations, the embryogenesis and the cell division.

Komamine: Yes, we looked for 21D7 antigen in proliferating and differentiating carrot cells in culture. We could detect it but these cells still have totipotency. If they were transferred to auxin-free medium, embryogenesis occurred. Unfortunately, we don't have a carrot suspension culture that has been subcultured for a long time and has lost totipotency.

Potrykus: Other laboratories may have cells suitable for these experiments.

van Montagu: Since you have an embryo-specific antigen, did you look for it in homologous systems? You mentioned periwinkle, did you see whether in other embryonic cultures this protein was present?

Komamine: Yes, Dr Jane Smith tried to detect the antigen, mainly in the intact plant. Meristematic tissue contains a lot of this antigen.

van Montagu: Did you look in other species to see if its structure has been conserved?

Komamine: It is also present in maize, peach, Arabidopsis and others.

Harada: Just one technical question. Is it difficult to select the small, round, cytoplasm-rich cells that you used for your experiment by physical or mechanical means rather than manually picking up each cell as you did?

Komamine: We tried by sieving or centrifugation. We used a Beckman centrifuge, Elutriator. We have not yet obtained reproducible results but it is possible to fractionate by this method. If you have any good ideas, please give me them.

Barz: Can you assign any enzymic activity to these embryogenesis-specific antigens?

Komamine: We have not yet looked, but what enzyme should we try to detect, there are so many enzyme activities?

Barz: Have you any explanation for the oxygen effect at the first stage of establishing the optimum conditions for the cells?

Komamine: Yes, a higher oxygen concentration (40%) promotes the induction of embryogenesis from single cells to an embryogenic cell cluster. However, usually we do not use this condition because the apparatus is so complicated. Ten or fifteen years ago it was reported that a lower concentration of oxygen promoted the induction of embryogenesis, but in that case the initial material may have been the cell cluster.

Fowler: That work was carried out with embryogenic suspension cultures of carrot cells. As he indicates, the embryogenesis occurred from cell clusters not individual cells. The objective of the work was to investigate the role of oxygen as a trigger for the onset of embryogenesis. Embryogenesis could be initiated in actively growing and dividing carrot cell suspension culture by reducing the dissolved oxygen concentration from about 40% saturation to about 10–12%. This resulted in a highly synchronized wave of embryogenesis throughout the culture. At the same time, major changes occurred in the levels of inorganic phosphate, AMP, ADP and ATP. We found that we could mimic the 'oxygen effect' by addition of nucleotide phosphates to the culture medium. Another interesting facet of this work was that it showed onset of embryogenesis to be accompanied by a major shift in the pattern of respiratory metabolism of the cells. Many plant cells have the potential to use a bypass system in the mitochondrial electron chain around phosphorylation Site III. (Interestingly, this bypass is also insensitive to cyanide, in contrast the more usual Site III/cytochrome a_3 system.) Onset of embryogenesis appears to be associated with a switch to the cyanide-insensitive pathway. We were unfortunately never able to determine whether the switch in respiratory metabolism occurred before or in response to the onset of embryogenesis.

Scowcroft: Professor Komamine, have you evaluated whether the 21D7 antibody interferes with normal zygotic embryogenesis?

Komamine: We could detect the presence of 21D7 antigen in zygotic embryos. We have not yet investigated whether injection of the antibody blocks zygotic embryogenesis. It is very difficult to microinject fertilized carrot eggs.

Scowcroft: In carrot that's certainly true, in other systems it may be more readily done.

Uchimiya: As I understand it, the 21D7 protein is a sort of DNA binding protein. Is that true?

Komamine: Yes, we detected the 21D7 protein only in the nuclei in cells, in State 0 single cells and State 1 and State 2 cell clusters. In every totipotent cell we could detect a signal of the 21D7 antigen only in the nucleus.

Zenk: In these very different stages of embryogenesis that you observed, did you ever see that carotenoid synthesis was turned on?

Komamine: We are interested in the expression of secondary metabolism during embryogenesis but we have not yet investigated this. We thought that we noticed a distinct smell in the cultures at a certain stage of embryogenesis. This may be due to the synthesis of particular stage-specific compounds.

Fowler: When reduction in the level of dissolved oxygen is used to initiate embryogenesis in carrot cell cultures, there is typically a massive increase in synthesis of fatty acids, erusic acid being a key component. Quite what is the significance of this activity in the specific context of the onset of embryogenesis, I am not sure.

Yamada: Have you ever tried to inject the 21D7 protein itself into the cytoplasm?

Komamine: We have not yet isolated the protein. We intend to isolate cDNAs for messenger RNAs specifically expressed during embryogenesis and we are also trying to isolate the 21D7 protein. We would like to microinject such a high molecular weight compound into elongated cells which seem to have lost totipotency. If the compound is an inducer of embryogenesis, then the elongated cells may divide and differentiate.

Takebe: You suggested that one of the a,b,c proteins that you see in 2-D gels might be the antigen recognized by the monoclonal antibody for the embryogenesis-specific protein.

Komamine: Protein A is the same molecular weight as the 21D7 protein, 45 kDa.

Takebe: Can you test that directly by Western blotting of the 2-D gels?

Komamine: We are trying. The result appears to be positive but on the 2-D gels it is hard to get a good signal.

References

Neuhaus G, Spangenberg G, Mittelsten-Scheid O, Schweiger HG 1987 Transgenic rapeseed plants obtained by the microinjection of DNA into microspore-derived embryoids. Theor Appl Genet 74:30–36

Protoplast fusion-mediated transfer of male sterility and other plasmone-controlled traits

Esra Galun, Avihai Perl and Dvora Aviv

Department of Plant Genetics, The Weizmann Institute of Science, Rehovot 76100, Israel

Abstract. Genetic and biochemical evidence indicates that cytoplasmic male sterility (CMS) in Gramineae and Solanaceae crop species is controlled by nuclear genome–chondriome (mitochondrial genome) interactions. This evidence is especially strong for maize, tobacco and petunia. We introduced the chondriome or some of its components from an alien donor species into a target cultivar by the donor–recipient protoplast fusion procedure and showed that CMS can be induced in a given tobacco cultivar by organelle transfer. This procedure has been used successfully in several laboratories to combine CMS with plastome-controlled atrazine tolerance in cultivars of *Brassica*, in order to establish desirable 'seed parents' for F1 hybrid seed production. In recent studies with *Nicotiana* we investigated pre-fusion treatments and plastome mutants as means by which to control organelle transfer from donor protoplasts to target plants. Such treatments have been used to produce male sterile potato cultivars. CMS cultivars should be very useful as seed parents for the production of F1 hybrid true potato seeds, which at present are produced by laborious manual emasculation and pollination.

1988 Applications of plant cell and tissue culture. Wiley, Chichester (Ciba Foundation Symposium 137) p 97–112

The coding capacity of plant cell organelles

Our knowledge of the coding capacity of the chloroplast genome (plastome) of angiosperms was vastly increased by the deciphering of the complete nucleotide sequence of the *Nicotiana* chloroplast (ct) DNA (Shinozaki et al 1986). On the basis of these and other studies, and the conservation of ctDNA among higher plants, the plastome is estimated to code for more than 40 polypeptides and contains many open reading frames. On the other hand, for only a few specific traits is there molecular evidence for a correlation between specific base sequences of ctDNA and the expression of these traits; for example, resistance to atrazine (see below), pigmentation deficiency (e.g.

97

Fluhr et al 1985) and resistance to certain antibiotics (e.g. Fromm et al 1987). Evidently, a procedure that enables the transfer of presumptive mutated chloroplasts (but neither the mitochondria nor nuclear components) from the plant harbouring this mutation into the cells of another plant will be an effective means of establishing such a correlation. Such a procedure has been developed and is described below.

We know that the coding capacity of the mitochondrial genome (chondriome) of angiosperms is much smaller than that of the plastome. The maize chondriome codes for all ribosomal RNAs and probably for all the transfer RNAs required for protein synthesis in the mitochondrion. It also codes for at least two cytochrome c oxidase subunits (I and II), for the apo-cytochrome b, for the α subunit of the F_1 ATPase complex, and for at least two subunits (6 and 9) of the F_0 ATPase complex (see Eckenrode & Levings 1986). This coding capacity seems to be shared by the chondriomes of all angiosperms, although evidence to substantiate this assumption is still lacking.

In spite of this limited known coding capacity of the chondriome there is convincing evidence that the chondriome participates in the control of important breeding traits. One of these, cytoplasmic male sterility (see Hanson & Conde 1985, Pring & Lonsdale 1985), is discussed below.

Organelle transfer and cybrids

As noted above, organelle transfer could be an efficient tool for the establishment of a correlation between the organelle genome and a specific trait. Moreover, if a given trait is relevant to breeding, such as herbicide resistance (of chloroplasts) or cytoplasmic male sterility (CMS, attributed to the chondriome), organelle transfer should be applicable to crop improvement.

Several years ago we developed a protoplast fusion procedure which could be used for organelle transfer among plants (Zelcer et al 1978, see Galun & Aviv 1986 for later references). We called this procedure donor–recipient protoplast fusion. Briefly, it involves the fusion of protoplasts in which nuclear division has been arrested by X or gamma-irradiation, with protoplasts which have been transiently exposed to a metabolic inhibitor, e.g. iodoacetate. The fused heteroprotoplasts are then plated, usually among gamma-irradiated feeder protoplasts, in an appropriate culture medium. The resulting colonies are regenerated into plants. In this fusion procedure the division-arrested protoplasts serve as organelle donors, while the non-irradiated protoplasts act as recipients of these organelles. Results from our laboratory, as well as studies by other authors, indicated that the resulting plants are commonly cybrids, i.e. contain organelles from both fusion partners but the nuclear genome from only one. In most cases, cybrids contain chloroplasts of either the donor or the recipient fusion partner; but heteroplastomic cybrids were recorded, which may transmit the heteroplastomic state to their sexual progeny.

The fate of mitochondria differs from that of chloroplasts. Cybrids usually contain mitochondria with novel mtDNA restriction patterns, indicating that recombination between the chondriomes of the fusion partners has taken place. It is important to note that the plastomes' segregation among the cybrids derived from one fusion event is independent of the segregation of chondriome-controlled traits, e.g. CMS. Note also that while this procedure causes the transfer of intact chloroplasts from a donor to a recipient plant, the mitochondria are not transferred as intact organelles. On the other hand, chondriome-coded traits are transferred. Thus, for brevity, we may speak about 'mitochondrial transfer' by the donor–recipient protoplast fusion procedure. The results of our studies on the use of this process for crop improvement have been reported recently (Galun et al 1987, Galun & Aviv 1987).

A glance at cytoplasmic male sterility

Cytoplasmic male sterility indicates a lack of functional pollen with retention of female fertility and is transmitted during sexual propagation together with cytoplasmic organelles. The histological, morphological and biochemical aspects of CMS in flowering plants were reviewed by Bino (1986). CMS has a very wide range of morphological manifestations, from a complete lack of stamens, through impairment of meiosis in the sporogenic tissue and arrest at the tetrad stage, to the production of mature but infertile pollen grains. Do these phenomena, reported in more than 170 species of about 20 plant families, have a common denominator? Almost all published studies implicate mitochondria in CMS. Before we describe our own studies on organelle transfer and CMS in *Nicotiana* and *Solanum*, let us briefly look at *Zea*, *Petunia* and *Brassica*.

Mitochondria have been implicated in CMS in maize by a great number of studies (see Pring & Lonsdale 1985). Two laboratories have reported detailed molecular evidence which correlates the T-cytoplasmic male sterility with a frame-shift in mitochondrial open reading frames. This shift may well alter the expression of such reading frames and thus cause CMS (Wise et al 1987 and references therein). In spite of the vast amount of circumstantial evidence, we do not have direct proof that there are 'CMS mitochondria' in maize. Unfortunately, cell manipulation methods which would enable the specific transfer of 'CMS mitochondria' from a CMS plant to a fertile plant, thereby showing that such mitochondria do confer CMS, are not available.

The sexual and non-sexual transmission, as well as the histology and molecular biology of CMS in *Petunia*, have been studied in detail, especially by R.J. Bino, E.C. Cocking, R. Frankel, S. Izhar, M.R. Hanson and their collaborators (see Frankel & Galun 1977, Young & Hanson 1987): only one CMS system has been described. Somatic hybridizations indicated that CMS did not co-segregate with the chloroplasts but was correlated with a mtDNA fragment. The chondriome of CMS *Petunia* cybrids contains aberrantly integrated *atp* 9 and *cox II* genes in addition to the normal homologues. It was found that the integrated gene is expressed in CMS plants and that the expression is developmentally regulated, showing an increase in the reproductive tissues. Whether the respective aberrant polypeptides are incorporated into the mitochondrial F_0 ATPase and/or cytochrome *c* oxidase, and impair the functioning of these enzymes, is not yet known.

Indications that mitochondria are involved in CMS in *Brassica* came primarily from breeding and somatic hybridization studies (Galun & Aviv 1987). Pelletier et al (1983) derived a somatic hybrid plant from the fusion of protoplasts of a male fertile and triazine-sensitive *B. campestris* line with protoplasts of a CMS line of *B. napus*. The somatic hybrid plant was triazine tolerant and showed CMS. This excludes the possibility that the 'cytoplasmic' male sterility was conferred by the triazine-tolerant chloroplasts. Similar results were obtained by Robertson et al (1987). Menczel and collaborators, who were using our donor–recipient protoplast fusion process in their *Nicotiana* studies, applied the same process to *Brassica* (Menczel et al 1987). They found that CMS could be transferred among *B. napus* lines and that this transfer was correlated with novel mtDNA restriction patterns. Finally, Barsby et al (1987) combined the donor-recipient protoplast fusion process with metabolic inhibition of the recipient protoplasts. They fused CMS donor protoplasts with recipient protoplasts which contained triazine-resistant chloroplasts and thereby transferred CMS to *B. napus*-like cybrids.

Studies with plants of several other genera, e.g. *Tricitum, Sorghum, Beta*, also indicated that mitochondrial–nuclear interactions are involved in CMS; a detailed account of these studies is beyond the scope of this paper.

Organelle transfer and CMS in Nicotiana

As indicated above, we found in our early cybridization studies (Zelcer et al 1978, Galun et al 1982) that the protoplast fusion-mediated transfer of CMS from a donor to *N. sylvestris* is correlated with the transfer of mitochondrial components, and that the male fertility of a CMS recipient can be restored by fusion with protoplasts of a fertile donor (Aviv & Galun 1980). Recently, we reported that the fusion of protoplasts from two different CMS cybrids may result in fertile plants (Aviv & Galun 1986). The restriction patterns of mtDNAs in the fertile plants are rearranged relative to those of both fusion partners. Such rearrangements are common in protoplast fusion progeny but it should be noted that rearrangement of mtDNA *per se* does not necessarily lead to changes in male fertility.

We then asked whether in such a cybrid, with restored fertility, the nuclear genome underwent a change during the course of fusion, culture and regeneration, causing a nuclear-coded restoration of fertility   We used a fertility-restored derivative of cybrid G-2-1-1-k/8 as a recipient and fused its protoplasts with protoplasts of a CMS donor, $92str^R$-7, spe^R-5. Before fusion, the recipient protoplasts were treated with 0.3 mM or 0.5 mM iodoacetate and the donor protoplasts were X-irradiated. The resulting calli were regenerated in the presence or absence of streptomycin and their male fertility was evaluated (Table 1). The results indicated that the iodoacetate treatment reduced the proportion of the streptomycin-sensitive recipients' chloroplasts in the fusion derivatives. Thus most cybrids had str^R chloroplasts. Furthermore, there was a high ratio of co-transfer of such chloroplasts with CMS, only a few cybrids had str^S chloroplasts and showed CMS (one cybrid which regenerated in the presence of streptomycin turned out to be str^S). After pre-fusion treatment with 0.5 mM iodoacetate, all cybrids which regenerated in the presence of streptomycin were CMS and str^R, indicating that both the chloroplasts and the chondriome components involved in CMS were transferred from the donor to the cybrids. Since some of the cybrids were sterile in spite of their G-2-1-1-k/8 nucleus, the latter did not contain a gene capable of restoring fertility.

In other experiments we tested the effect of pre-fusion treatment with rhodamine 6G (R6G). This is a lipophilic dye which binds to mammalian mitochondrial membranes and impairs their function; mammalian cells that were treated with R6G and subsequently fused with other cells did not contribute their mitochondria to the resulting heterokaryons. We investigated the effect of R6G on plant protoplasts used as donor–recipient fusion partners (Aviv et al 1986). We found that cybrid *Nicotiana* plants can be produced efficiently when X-irradiated donor protoplasts are fused with R6G-treated recipient protoplasts. The latter usually did not divide unless fused with the donor protoplasts; the donor protoplasts did not produce colonies because nuclear division was arrested. Moreover, when X-irradiated protoplasts of *N. rustica* were fused with R6G protoplasts of the double mutant SR1 pig^{-35} (streptomycin resistant, albino), chondriome analysis of the cybrids indicated that cybrids had only one of two chondriome compositions. We did not detect rearranged mtDNA in the cybrids. Such an 'intact' transfer of the chondriome could be specific to fusions involving *N. rustica*. We therefore did another donor–recipient fusion in which SR1 pig^{-35} protoplasts were treated with R6G before fusion but then fused with $92 lin^R$-17 donor protoplasts containing *N. undulata* mtDNA. This fusion resulted in rearranged mtDNA in the cybrid derivatives. Only one cybrid, out of 12 that were analysed, contained the donor's (i.e. *N. undulata*) mtDNA. The other 11 cybrids contained novel mtDNA.

The effect of R6G on the mitochondria of treated recipient protoplasts could be dose dependent. We therefore performed extensive fusion experiments in which the donor protoplasts were X-irradiated as well as being treated with various doses of R6G. The recipient protoplasts were treated before fusion with 0.5 mM iodoacetate. $92 lin^R$-17 (CMS; mitochondria and

TABLE 1   Cybrid progeny from the fusion between 92 str^R-7, spe^R-5 (CMS, streptomycin and spectinomycin resistant) donor protoplasts and iodoacetate-treated G-2-1-1-k/8 (restored fertility, streptomycin sensitive) recipient protoplasts

Iodoacetate (mM)	Cybrid Male fertile or sterile	Progeny $+Str^a$	Classification $-Str^b$
0.3	Fertile	None	None
	Part fertile	2/2	0/2
	Sterile	4/5	2/4
0.5	Fertile	None	1/1
	Part fertile	None	None
	Sterile	12/12	10/13

[a] Number of streptomycin-resistant cybrids/number of total cybrids regenerated on streptomycin.
[b] Number of streptomycin-resistant cybrids/number of total cybrids regenerated without streptomycin.

chloroplasts of *N. undulata*) and SR1 *pig*$^{-35}$ (fertile; mitochondria and chloroplasts of *N. tabacum*) served as donor and recipient, respectively. Only green cybrids were retained, which all contained the plastome of the CMS donor (the recipient had a plastome mutation causing chlorophyll deficiency). Out of 28 cybrids in which the mtDNA was analysed, novel or rearranged chondriomes were found in 24 plants. Fusions in which the donor protoplasts were treated with 10, 15 or 20 µg/ml R6G resulted in both fertile cybrids and cybrids showing CMS. The fusion, which involved treating donor protoplasts with 25 µg/ml R6G, produced six cybrids. Four of these showed CMS and had rearranged chondriomes, while two plants were fertile (as the recipient) and their mtDNA restriction profiles (as revealed by Southern blot hybridization with three probes) were indentical to the corresponding profiles of the recipient, *N. tabacum*. The latter two plants were cybrids because they contained the donor's chloroplasts. It thus seems that a very high dose of R6G could eliminate the chondriomes in the treated donor protoplasts, while the chloroplasts were retained, at least in part, and transferred to the recipient. Further tests are required to substantiate these results.

Protoplast fusion and CMS in potato

Potato protoplasts were used in the pioneering work of Melchers and his collaborators in the first attempt to produce inter-generic somatic hybrids (of potato and tomato). Such protoplasts were also used by several investigators, including H. Binding, G.S. Bokelman, J. Gressel, E. Shahin, E. Thomas, G. Wenczel and their collaborators, to regenerate plants and in various fusion experiments (see Miller & Lipschutz 1984). Helgeson and his collabor-

TABLE 2 Donor-recipient protoplast fusions between a CMS donor or wild Solanum species and potato cultivars

| Fusion partners | | | | |
Donor	Recipient	Aim of fusion	Number of calli obtained	Stage of experiment[a]
Y245.7 (CMS)	S. tuberosum cv. Atzimba	Transfer of CMS	21	Flowering cybrids
Y245.7 (CMS)	S. tuberosum cv. Atlantic	Transfer of CMS	2	Flowering cybrids
Y245.7 (CMS)	S. tuberosum cv. Desiree	Nuclear/cytoplasmic compatability	(Many)	Calli in liquid medium
S. demissum	S. tuberosum cv. Desiree	Nuclear/cytoplasmic compatability	55	Mature cybrid plants
S. brevidens	S. tuberosum cv. Desiree	Nuclear/cytoplasmic compatability	6	Regenerated cybrids
S. chacoense	S. tuberosum cv. Desiree	Nuclear/cytoplasmic compatability	10	Regenerated cybrids
S. commersonii	S. tuberosum cv. Desiree	Nuclear/cytoplasmic compatability	7	Regenerated cybrids
S. etuberosum	S. tuberosum cv. Desiree	Nuclear/cytoplasmic compatability	58	Regenerated cybrids
S. berthaultii	S. tuberosum cv. Desiree	Nuclear/cytoplasmic compatability	24	Regenerated cybrids
S. nigrum	S. tuberosum cv. Desiree	Nuclear/cytoplasmic compatability	(Many)	Calli in liquid medium

[a] As of July 1987.

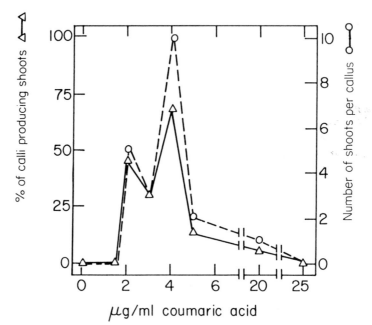

FIG. 1.   Promotion of shoot regeneration from cybrid potato calli. Cybrid calli were obtained by fusing gamma-irradiated *Solanum berthaultii* (donor) protoplasts and iodoacetate-treated protoplasts of the potato cultivar Desiree (recipient). Shoots were scored 21 days after transfer of microcalli to regeneration medium containing different levels of *o*-coumaric acid. Each point represents the mean of 22–34 calli.

ators initiated investigations designed to produce inter-specific somatic hybrids between potato and wild *Solanum* species in order to transfer disease resistance from such species into potato cultivars (Helgeson et al 1986).

   Our interest in organelle transfer in potato was initiated by a collaboration with the International Potato Center (CIP) in Lima, Peru. CIP is developing F1 hybrid true potato seeds for farmers in developing countries to use as a substitute for seed tubers. Our aim was to eliminate the laborious manual emasculation involved in producing hybrid true potato seeds. We thus started donor–recipient protoplast fusions in order to transfer CMS from a donor line into cultivars selected by CIP as prospective seed parents (e.g. 'Atzimba' and 'Atlantic'). Cybrids have been obtained which indicate that we have probably successfully transferred the CMS trait. Several cybrid plants had non-germinating pollen: some had shrunken pollen or pollen arrested at the tetrad stage. Cybrids of one callus, derived from the fusion between protoplasts of the CMS source (Y245.7) and 'Atzimba' protoplasts, produced stamens which were devoid of pollen and had petaloid anthers. All the cybrids analysed had novel mtDNA.

   Our present interest is to use the vast range of *Solanum* species to investigate plastome–nuclear genome and chondriome–nuclear genome compatibilities. We have thus started donor–recipient fusions between potato and several *Solanum* species (Table 2). In the course of these fusion experiments we encountered, in some fusion combinations, difficulties in the regeneration of plants from cybrid calli. Application of *o*-coumaric acid seemed to overcome this difficulty (Fig. 1).

A lesson from yeast: is it applicable to tobacco?

Yeast mitochondrial ATPase, like that of higher plants, is composed of F_1 and F_0 components. Some of the subunits of F_0 are encoded by the nuclear genome and some by the chondriome. A *Saccharomyces cerevisiae* strain, 990, was isolated that had a mutation in the mitochondrial gene *oli*1, which codes for subunit 9 of ATPase. This is a 'conditioned' mutant: it will not grow on glycerol at 20 °C but will grow on non-fermentable carbon sources at 28 °C. When cultured at 28 °C it is also oligomycin resistant. Notably, out of 10 revertants of this mutated line, two were attributed to nuclear events (suppressor genes) and eight to the mitochondria. In seven of these eight revertants the base sequences reverted to normal, whereas in one there was an additional base sequence change (Hefta et al 1987).

We recently started to isolate mitochondrial mutants in *Nicotiana*. An *N. sylvestris* cell suspension was exposed repeatedly to antibiotics to select for resistant cell lines. Several lines with resistance to oligomycin (*oli*^{R}) were isolated. One of these lines, *N. sylvestris oli*^{R}-38, was chosen for further studies, mainly to establish whether or not the oligomycin resistance of this line is encoded by the mitochondrial genome and, if it is, whether or not the mutation involves oligomycin-resistant ATPase (i.e. subunit 9 of the F_0 ATPase). To answer these questions we used two approaches. In the first, protoplasts of *oli*^{R}-38 were fused as donors with protoplasts of the CMS line 92 (*N. tabacum* nuclei, plastome and chondriome of *N. undulata*). We knew from previous studies that such a CMS line can be made fertile by transfer of mitochondria. Thus, fertile cybrids are expected from this fusion. Will some of these cybrids be oligomycin resistant? We recently obtained such fertile cybrids and are currently analysing their plastomes. If these cybrids have *N. undulata* plastomes but express oligomycin resistance, there will be a strong indication that *oli*^{R}-38 is indeed a mitochondrial mutation. This will be the first mitochondrial mutant induced in higher plants. We are also looking for evidence of the involvement of mitochondria in the oligomycin resistance by studying, '*in organello*', the ATP-synthase activity in the presence and absence of oligomycin, using either *oli*^{R}-38 or control mitochondria.

Acknowledgements

Our research was supported by an AID grant (No DFE 5542-G-55-4030), by a grant from the National Council for Research and Development (Israel) and GSF (Munich, FRG) and by the Julia Forschheimer Center for Molecular Genetics. Esra Galun is an incumbent of the Irene and David Schwartz Chair for Plant Genetics.

References

Aviv D, Chen R, Galun E 1986 Does pretreatment by rhodamine-6-G affect the mitochondrial composition of fusion-derived *Nicotiana* cybrids? Plant Cell Rep 5: 227–230

Aviv D, Galun E 1980 Restoration of fertility in cytoplasmic male-sterile (CMS) *Nicotiana sylvestris* by fusion with X-irradiated *N. tabacum* protoplasts. Theor Appl Genet 58:121–127

Aviv D, Galun E 1986 Restoration of male fertile *Nicotiana* by fusion of protoplasts derived from two different cytoplasmic-male-sterile cybrids. Plant Mol Biol 7:411–417

Barsby TL, Chuong PV, Yarrow SA et al 1987 The combination of Polima CMS and cytoplasmic triazine resistance in *Brassica napus*. Theor Appl Genet 73:809–814

Bino RJ 1986 Cytoplasmic male sterility in *Petunia* hybrida. PhD Thesis, Agricultural University, Wageningen

Eckenrode VK, Levings III CS 1986 Maize mitochondrial genes. In Vitro Cell & Develop Biol 22:169–176

Fluhr R, Aviv D, Galun E, Edelman M 1985 Efficient induction and selection of chloroplast encoded antibiotic-resistant mutants in *Nicotiana*. Proc Natl Acad Sci (USA) 82:1485–1489

Frankel R, Galun E 1977 Pollination mechanisms, reproduction and plant breeding. Springer Verlag, Berlin p 281

Fromm H, Edelman M, Aviv D, Galun E 1987 The molecular basis for rRNA dependent spectinomycin resistance in *Nicotiana* chloroplasts. Eur Mol Biol Organ (EMBO) J 6:3233–3239

Galun E, Arzee-Gonen P, Fluhr R, Edelman M, Aviv D 1982 Cytoplasmic hybridization in *Nicotiana*: mitochondrial analysis in progenies resulting from fusion between protoplasts having different organelle constitutions. Mol Gen Genet 186:50–56

Galun E, Aviv D 1986 Organelle transfer. Methods Enzymol 118:595–611

Galun E, Aviv D 1987 Fundamental and applied aspects of plant-organelle transfer by protoplast manipulation. National University, Bogota (Proc Int Congress Plant Tissue Culture-Tropical Species, Bogota, Columbia 1987), in press

Galun E, Aviv D, Breiman A, Fromm H, Perl A, Vardi A 1987 Cybrids in *Nicotiana*, *Solanum* and *Citrus*: isolation and characterization of plastome mutants: pre-fusion treatments, selection and analysis of cybrids. NATO Adv Study Inst, Copenhagen 1987. In: von Wettstein D, Chua NH (eds) Plant Molecular Biology. Plenum, New York p 199–207

Hanson MR, Conde MF 1985 Functioning and variation of cytoplasmic genomes: lessons from cytoplasmic-nuclear interactions conferring male sterilities in plants. Int Rev Cytol 94:213–265

Hefta LJF, Lewin AS, Daignan-Fornier B, Bolotin-Fukuhara M 1987 Nuclear and mitochondrial revertants of a mitochondrial mutant with a defect in the ATP synthase complex. Mol Gen Genet 207:106–113

Helgeson JP, Hunt GJ, Habelach GT, Austin S 1986 Somatic hybrids between *Solanum brevidens* and *Solanum tuberosum*: expression of a late blight resistance gene and potato leaf roll resistance. Plant Cell Rep 5:212–214

Menczel L, Morgan A, Brown S, Maliga P 1987 Fusion-mediated combination of

Ogura-type cytoplasmic male sterility with *Brassica napus* plastids using X-irradiated CMS protoplasts. Plant Cell Rep 6:98–101

Miller SA, Lipschutz L 1984 Potato. In: Ammirato PV et al (eds) Handbook of plant cell culture. Macmillan, New York vol 3:291–326

Pelletier G, Primard C, Vedel F et al 1983 Intergeneric cytoplasmic hybridization in Cruciferae by protoplast fusion. Mol Gen Genet 191:244–250

Pring DR, Lonsdale MD 1985 Molecular biology of higher plant mitochondrial DNA. Int Rev Cytol 97:1–46

Robertson D, Palmer JD, Earle ED, Mutschler MA 1987 Analysis of organelle genomes in a somatic hybrid derived from cytoplasmic male-sterile *Brassica oleracea* and atrazine-resistant *B. campestris*. Theor Appl Genet 74:303–310

Shinozaki K, Ohme T, Tanaka M et al 1986 The complete nucleotide sequence of the tobacco chloroplast genome: it's gene organization and expression. Eur Mol Biol Organ (EMBO) J 5:2043–2049

Wise RP, Pring DR, Gengenbach BG 1987 Mutation to male fertility and toxin insensitivity in Texas (T)-cytoplasm maize is associated with a frameshift in a mitochondrial open reading frame. Proc Natl Acad Sci (USA) 84:2858–2862

Young EG, Hanson MR 1987 A fused mitochondrial gene associated with cytoplasmic male sterility is developmentally regulated. Cell 50:41–49

Zelcer A, Aviv D, Galun E 1978 Interspecific transfer of cytoplasmic male sterility by fusion between protoplasts of normal *Nicotiana sylvestris* and X-ray irradiated protoplasts of male-sterile *N. tabacum*. Z Pflanzenphysiol 90:397–407

DISCUSSION

Scowcroft: Professor Galun, you have demonstrated with great clarity that this technology of 'cybridization' is now a workable tool in the solanaceous crops. We are interested in using this in the *Brassica* species. It has always intrigued me, at least from a theoretical standpoint, that there is a biological contradiction in cybridization. You are using X-rays to destroy the nucleus of the donor...

Galun: Not to 'destroy' but to cause division arrest in the irradiated nuclei.

Scowcroft: You macerate the genetic material to some extent?

Galun: To prevent it from further division.

Scowcroft: So you interfere with DNA replication, perhaps centromeric activity during mitosis. But there is some destruction of the chromatin?

Galun: You can still see the nucleus of the irradiated protoplast, you don't destroy it. Sometimes it even divides once or twice, but then it stops.

Scowcroft: Let me go on with the contradiction. Basically, cybridization provides a delivery system of DNA from your donor nucleus to the recipient. Why do you never see part of that DNA being incorporated into the recipient nucleus?

Galun: We see, rarely, some indications of influence of the 'donor's' nucleus in the phenotypes of the cybrids, but most, and in many cases all, of the cybrid progeny have the nuclear-coded characters of the 'recipient'.

Scowcroft: The second contradiction is that you are using X-rays; these will also affect the organelles, the mitochondria and the chloroplasts. How can you distinguish that effect of X-rays, the genetic effect on the chloroplasts and mitochondria, from recombination?

Galun: We actually tried to induce mutation in organelles by X-rays and gave up. You can irradiate them with up to 100 kRad and there is no effect. The reason is probably that the target number is enormous, especially if you are using mesophyll protoplasts, which may have something like 50000 chloroplast genomes. If you induce one or two mutations without selection you will never detect them.

Scowcroft: Another reason might be that the DNA of mitochondria and chloroplasts does not have any centromeric activity; that's where the consequences of X-irradiation occur.

Galun: X-irradiation is a very bad way to induce mutation. I don't know of any mutation which was induced in this way; it is very easy to induce mutations chemically. X-irradiation may cause the transfer of some nuclear characters into the cybrid, that has not been excluded. Some people even use that as a method, but with higher doses: they X-irradiate the nuclei then fuse them, but for that you need selective markers. For instance, if you have a nitrate reductase minus recipient, you X-irradiate the normal nucleus and then look for the ability of the fusion derivative to grow in nitrate. The number may be one in 10000, but it is possible to select. Normally, the amount of morphological change is minimal. We have thousands of cybrids, they all look like the recipient parents, very rarely do we see any change.

Potrykus: We have applied this donor-recipient technique and one can actually see nuclear DNA from the 'donor', if one uses DNA-specific probes, which is possible. By using species-specific DNA probes, and the technique Professor Galun described to produce clean cybrids, up to 80% of the donor nuclear DNA is in the supposed cybrids, so we think that one should not call this a cybrid but a hybrid (Imamura et al 1987).

Scowcroft: This has real consequences in a breeding programme because cybridization actually delivers large amounts of donor DNA to the recipient.

Potrykus: That's why I think that when interpreting the data, one should include the possibility that you have transferred massive amounts of nuclear genes.

Galun: I agree, it is a possibility.

Scowcroft: If you have that, if you then have to 'correct' the donor nucleus by a backcrossing programme, you may come unstuck in terms of the cytoplasmic organelles.

Galun: I think that you don't have to, because in our experience the cybrids have, with very little exception, exactly the same nuclear-coded features as the recipient.

Tabata: When you irradiate donor cells with X-rays, you may expect mutations to occur also in the mitochondria or plastid DNA. Isn't there a possibility that this might lead to another type of male sterility?

Galun: It is very difficult or even impossible to induce mitochondrial muta-tion. In our case, what is always correlated with male sterility is components of the donor's mitochondria. In petunia, Maurine Hanson and collaborators (Boeschore et al 1985) believe that they have exactly the fragments of DNA of the mitochondria which are the real cause of cytoplasmic male sterility; also after such transfer of mitochondria from one cell to another. So there is very little doubt that this is really the cause of male sterility. You can exclude completely the chloroplasts, at least in most of the Solanaceae, they have nothing to do with male sterility.

Cocking: Professor Galun has emphasized the important approach of using irradiation to inactivate the nucleus to aid the transfer system in that respect, but you can avoid using irradiation in the production of cybrids by using methods, now pretty well established, to produce enucleate material. So you could do some comparative assessments, in which you did not use any irradia-tion at all but used enucleated protoplasts for your fusion procedures. There have been one or two papers published along those lines (Maliga et al 1982, Bradley 1983); admittedly even they have sometimes been criticized for not 100% enucleation of protoplasts. I think the methodology is improving and one could do very useful comparative assessments in that connection.

Riley: I greatly admire the permutations of organelles and nuclei that you have described. I wonder whether you had considered another problem in potato breeding. Many potato cultivars are already male sterile. There is therefore the problem for potato breeders that some hybrid combinations of cultivars cannot be made because both potential parents are male sterile. There is now apparently a means of making potato cultivars male fertile where they are not naturally so.

Galun: I didn't go into the breeding part here. Actually the breeding compo-nent is handled by the breeders at the International Potato Center (CIP) in Lima, Peru. They have already selected pairs of parents which work very nicely by hand pollination, that means they have good pollinators and they have good seed parent cultivars, only the seed parent cultivar is male fertile. They want to avoid emasculation, therefore they ask us to eliminate the anthers or the pollen from this cultivar. We are not involved in potato breeding ourselves, it is too complicated, we just did a task.

Riley: I'm not suggesting that you should do it, I am saying that there is a way by which it could be done.

Harms: In the case where you have male fertile and male sterile plants from the same callus, or plants which represent different plastome constitutions, do you suggest that the callus is chimaeric to start with or that in the course of regeneration a rapid sorting of parental organelles occurs to give the result that you obtain?

Galun: You start with a heteroplasmonic fused cell which contains a mixed population of chloroplasts and nuclei from both fusion partners. Then during cell division, there is a gradual sorting out. In the same callus you may have cells which started with different types of organelles. Each of those cells may start to generate a plant, so you may have from the same callus different types of plants having different organelle composition.

Scowcroft: Has there been any progress in fusion of protoplasts with isolated chloroplasts or mitochondria, perhaps encapsulated in artificial but biologically active membranes?

Galun: I don't know and I don't believe there will be.

Scowcroft: Why don't you believe so?

Galun: Those organelles are quite sensitive and when you isolate them and purify them, I think that by the time you have finished the purification they will be effectively dead with respect to being capable of further replication. I think there is a chance to penetrate a living protoplast with a micropipette, take out organelles and immediately inject them into another protoplast. The system which has been used up till now, to first purify very meticulously and be sure that you don't have any other components in your suspension, and then inject and hope that the organelle has survived, has not worked and I don't think that it will.

Ohyama: Concerning the transfer of organelles to different cells, I don't think that it is impossible to do. When you do this kind of experiment, you have to be careful because the composition of the genes in the chloroplasts from the different species is not the same. For example, the genes encoded in the liverwort chloroplast genome, as determined by our group, are not the same as those encoded by the chloroplast genome in tobacco, as determined by Dr Sugiura. Therefore, if you transfer chloroplasts from cells of one species to those of another, the chloroplast genes do not necessarily complement the nuclear-encoded genes in the recipient cells.

Potrykus: There is a technique established now which offers an alternative method to the direct transfer of organelles. This technique combines microculture of isolated protoplasts in nanolitre droplets of culture medium with electrofusion. Applying this system you can take controlled cytoplasts not containing a nucleus, and containing defined chloroplasts or defined cytoplasm, and fuse them one by one with karyoplasts or with total protoplasts. With this technique you can produce controlled combinations of cytoplasms and nuclei or complete protoplasts (Spangenberg et al 1986).

Galun: Was it carrried on to completion in plants?

Potrykus: It can be carried on to complete plants. By combining cytoplasts and karyoplasts, you can get a new combination. This is a very clean system because it transfers organelles without 'contaminating' nuclear DNA.

Scowcroft: This is using microspores?

Potrykus: In this case using protoplasts. The importance is that you work on the individual level in microdroplets of culture medium and you observe exactly what you are doing. So the criticism of sub-protoplast fusions which I always raise does not apply here because you can really control what you are doing, by applying electrofusion. The important point is the subsequent micro-culture of your defined fusion product in nanolitre droplets. As these dessicate very quickly, you have to culture the microculture droplet within a mineral oil droplet to keep it stable. Then you introduce your selected protoplast, cyto-plast, or karyoplast, into the volume of this microculture chamber. By applying electrofusion with microelectrodes, you can either fuse protoplasts or combine karyoplasts with a cytoplast to get cell reconsitution. Or you can combine meristematic protoplasts with terminally differentiated ones to do studies like those Ted Cocking indicated, e.g. the activation of a terminally differentiated protoplast via fusion with a cytoplast from a dividing protoplast.

Chaleff: Is there any transmission of organelles through the pollen?

Galun: One has to be specific for each case. In tobacco there are reports that there is a low percentage of transfer of mitochondria through the pollen. This is of little use for transfer of male sterility within a species, because this can be achieved by normal crossing. It is interesting, if you can induce male sterility in a species where it does not naturally occur, or the system is very simple. It's very rare to find transfer of male sterility by sexual crosses—first of all you cannot use usually the pollen of a male sterility source to pollinate because it doesn't work. So the first problem is how to transfer through the pollen, if the pollen is not fertile.

Chaleff: But there may be traits other than male sterility that one might want to transfer.

Galun: Yes, but not many of them are easy to recognize. If they are, they are usually connected with male sterility, like *Helmintosporium* resistance in maize. People did it because in several species this is almost the norm—in many species, about 10% of all angiosperms, there is biparental transfer of organelles. There are classical experiments in this, mainly in Germany, there were generations of geneticists, who used this as a system to investigate what happens after such transfer (e.g. Sears 1980).

Barz: Your work leads us to questions about the chemical analysis of cybrids. In your case it would be especially interesting to look at the lipids of the chloroplast. The pattern of lipids and the ratio of saturated to unsaturated fatty acids are quite often species specific. Have you looked at this?

Galun: I am sure there are people using the same system in *Brassica*. Our system is used by Barsby et al (1987) to combine male sterility and resistance to triazine and they did this successfully. There is also interest, of course, from people working with oil crops, who want to improve the oil composition of their seeds.

Barz: You don't know what the results are?

Galun: No, because they will tell us only after they finish their experiment.

Riley: Dr Yamada, could I turn this topic slightly and ask whether in organisms which have organelles, chloroplasts or mitochondria, derived from different parents there is such a thing as a heterotic response between these different genetic systems. Some time ago there was a great deal of speculation about what was called mitochondrial complementation, as a means of measuring the potential of a hybrid corn. Could mitochondrial complementation have validity?

Galun: The only clue is that many years ago Otto Schieder and others found that when they fused two types of protoplasts from two different *Datura* species by normal somatic hybridization, commonly the fused callus had a better multiplication rate than either of the parental types. This was a way by which one could select the hybrid calli from the non-hybrid calli. There are very few cases where you can keep intact two types of mitochondria. This doesn't work for a simple reason: in a different way from chloroplasts, mitochondria *in vivo* are not discrete organelles, they fuse and de-fuse. So once you bring two types of mitochondria into the same cell, sooner or later they will mix their DNA and there will be recombination of their DNA. So there is no case where you have two distinct types of mitochondria in the same cell for a long period. With chloroplasts there is this possibility, and by some means you can even artificially maintain it for one or two generations, even beyond the sexual cycle. But I wouldn't say that we have any indication that such a heteroplasmonic situation is more viable than the homoplasmonic state.

References

Barsby TL, Chuong PV, Yarrow SA et al 1987 The combination of Polima CMS and cytoplasmic triazine resistance in *Brassica napus*. Theor Appl Genet 73:809–814

Boeschore ML, Hanson MR, Izhar S 1985 A variant mitochondrial DNA arrangement specific to *Petunia* stable sterile somatic hybrids. Plant Mol Biol 4:125–132

Bradley PM 1983 The production of higher plant subprotoplasts. Plant Mol Biol Rep 1:117–123

Imamura J, Saul MW, Potrykus I 1987 X-Ray irradiation promoted asymmetric somatic hybridization and molecular analysis of the products. Theor Appl Gen 74:445–450

Maliga PH, Lörz H, Lazar G, Nagy F 1982 Cytoplast-protoplast fusion for interspecific chloroplast transfer in *Nicotiana*. Mol & Gen Genet 185:211–215

Sears BB 1980 Elimination of plastids during spermatogenesis and fertilization in the plant kingdom. Plasmid 4:233–255

Spangenberg G, Neuhaus G, Schweige HG 1986 Expression of foreign genes in a higher-plant cell after electro fusion-mediated cell reconstitution of a microinjected karyoplast and a cytoplast. Eur J Cell 42:236–238

Analysis of chloroplast genomes in parasexual hybrid calli

Atsushi Hirai, Shinji Akada* and Satomi Sugiura

Graduate Division of Biochemical Regulation, Faculty of Agriculture, Nagoya University, Nagoya, 464, Japan

Abstract. The mode of segregation of chloroplast genomes in parasexual hybrid calli and the ability of hybrid cells to differentiate shoots were studied using hybrids between *Nicotiana glauca* with *N. gossei* chloroplasts and the regular *N. langsdorffii.* This system was used to select hybrid cells by culture in a hormone-free medium and to identify chloroplast genomes by the isoelectric focusing patterns of the large subunit of Fraction I protein. Two kinds of chloroplast genomes seem to segregate rapidly. The proportion of *N. langsdorffii* chloroplasts varied from 17% to 92%, therefore that of *N. glauca* varied from 83% to 8%, within a two-month-old callus. Another experiment showed that 16% of cells in an equivalent callus still contained two kinds of chloroplasts. However, only 9.5% of shoots developed from an equivalent callus contained two kinds of chloroplast when total tissue grown from each shoot was analysed. We therefore tentatively propose that a cell containing two kinds of chloroplasts has less ability to differentiate and form a shoot than a normal cell.

1988 Applications of plant cell and tissue culture. Wiley, Chichester (Ciba Foundation Symposium 137) p 113–122

Chloroplast genomes contain important genetic information for photosynthesis. However, the majority of breeding programmes for crops paid relatively little attention to the chloroplast genomes until recently. This was because the appropriate technology needed to improve the genome, which is inherited solely through the maternal line, was lacking. One way to overcome these obstacles is to make parasexual hybrids so that two different types of chloroplasts are present in a single fused cell. This in turn enhances the possibility of inducing interspecific chloroplast hybridization. A recent report described an interspecific chloroplast recombination in a *Nicotiana* parasexual hybrid (Medgyesy et al 1985).

However, although some of the hybrid plants regenerated from fused cells have two types of chloroplasts, there have been many reports stating that the majority of the plants have only one type (Chen et al 1977, Schiller et al 1982, Glimelius et al 1981).

---

* *Present address:* Department of Botany, University of Maryland, College Park, Maryland, 20742 USA

We are interested in the mode of segregation of the chloroplast genomes in hybrid plants. In this paper, we describe how two different chloroplast genomes segregate during the development of a hybrid callus, and why the majority of hybrid plants have only one type of chloroplast.

The experimental system

The cell fusion between *Nicotiana glauca* and *N. langsdorffii* was conducted to obtain parasexual hybrids, in order to select the hybrids by culture in a hormone-free medium (Carlson et al 1972), while the calli were still very small.

We used the large subunit (LS) of the Fraction I protein (ribulose-1, 5-bisphosphate carboxylase oxygenase), encoded by the chloroplast genome (Chan & Wildman 1972), as a chloroplast genetic marker, and the small subunit (SS), encoded by the nuclear genome (Kawashima & Wildman 1972), for the identification of hybrids. The LS bands from *N. glauca* are focused at higher pH than those from *N. langsdorffii* in an isoelectric focusing gel containing 8M urea. However, it is difficult to distinguish the two. The LS bands from *N. gossei* are focused at still higher pH than those from *N. glauca* when the Fraction I proteins are carboxymethylated. Therefore we obtained *N. glauca* with *N. gossei* chloroplasts by crossing female *N. gossei* and male *N. glauca*, then backcrossing with *N. glauca* (Akada & Hirai 1986b). Cells from this plant were fused with *N. langsdorffii* cells, and hybrid calli were selected in hormone-free medium.

Chloroplasts in the hybrid callus

Since most of parasexual hybrid plants contained only one of the two chloroplast genomes, it would be interesting to know when the elimination of one

TABLE 1   Kind of chloroplast and nuclear genomes in individual somatic hybrid calli

Type and species of cells used for fusion of protoplasts		Chloroplast genomes			Nuclear genomes		
N.glauca (G)	*N.langsdorffii* (L)	(No. of calli)			(No. of calli)		
		G	G+L	L	G	G+L	L
Suspension cells	Suspension cells	4	9	0	0	13	0
Suspension cells	Leaf cells	1	6	3	0	10	0
Leaf cells	Suspension cells	5	5	1	0	11	0

kind of chloroplast occurred. We therefore analysed the chloroplast genomes in hybrid calli between *N. glauca* and *N. langsdorffii*. As shown in Table 1, the results from calli were different from those obtained from plants. However, after serial subculture of those original hybrid calli, we found predominantly calli containing only one type of chloroplast (Akada et al 1983). Further evidence came from the *Bam*HI restriction pattern of chloroplast DNA as a chloroplast marker (Ichikawa et al 1984). Furthermore, we found that the ratio of the two kinds of chloroplast genomes differed within a callus (Akada & Hirai 1983). These results suggested that even the original hybrid calli contained two types of chloroplasts, i.e. they were heterogenous at the chloroplast level.

Chloroplasts in each of the eight sub-divisions of the hybrid callus

Two of the two-month-old hybrid calli between *N. glauca* with *N. gossei* chloroplasts and *N. langsdorffii* were cut into eight parts. Each part was subcultured for one month and then subjected to analysis of its Fraction I protein. Again, two types of LS were detected from all of the eight parts. We measured the intensity of the main bands from *N. gossei* and *N. langsdorffii* by densitometry. As shown in Fig. 1, the ratio of the two types of chloroplasts in each of the sub-divisions is entirely different (Akada & Hirai 1986a).

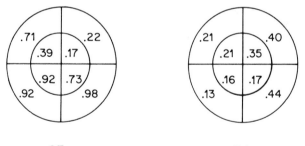

.63   .24

FIG. 1.   Distribution of the two types of chloroplasts in parasexual hybrid calli. Two-month-old parasexual hybrid calli were sub-divided into eight parts and the chloroplast types in each part analysed by isoelectric focusing of the Fraction I protein. The inner areas of the circles represent the upper part of the calli and the outer areas represent the lower part. The numbers shown are the the proportions of the *N. langsdorffii* kind of LS in the total LS (*N. langsdorffii* + *N. gossei*) (Akada & Hirai 1986a).

Chloroplasts in single cells of the hybrid callus

Small pieces from two-month-old hybrid callus were subcultured for four more months. Because four months is enough for four grams of callus to be regenerated from a single cell, it can be assumed that a callus of that size after

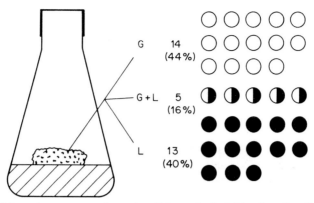

FIG. 2. Chloroplast genomes in each cell from the hybrid callus. Small pieces from the original hybrid callus were subcultured for four months. Four grams of the tissue from each piece, which was estimated by the growth rate to have originated from a single cell of the original callus, were used for determination of chloroplast genome type by Fraction I protein analysis. (From Akada & Hirai 1986a.)

four months of culture has originated from a single cell. These single cell-equivalent calli were used for Fraction I protein analysis. The results are illustrated in Fig. 2: 84% of the single cell-equivalent calli have only one parental type of chloroplast; 16% have two types of chloroplast. These results support the random segregation of chloroplasts during the development of a callus (Akada & Hirai 1986a).

Chloroplasts in small shoots developed from the hybrid callus

Small shoots about 2 mm long from the original two-month-old calli were

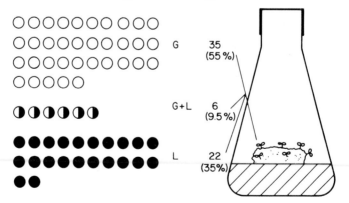

FIG. 3. Chloroplast genomes in each shoot from the hybrid callus. Small shoots differentiated from the original hybrid callus were subcultured until they weighed about two grams, then the chloroplast genome type was determined by analysis of the Fraction I proteins.

transferred to new flasks containing agar medium, and cultured until each weighed about two grams. Each tissue was analysed for the type of Fraction I protein in the chloroplast genomes. As shown in Fig. 3, 35 shoots contained only *N. gossei* type chloroplasts and 22 shoots contained *N. langsdorffii* type. Six shoots, corresponding to 9.5% of the total number of shoots analysed, contained two parental types of chloroplasts.

Discussion

On the basis of the available information, we concluded that the two types of chloroplast segregate at random during the development of the parasexual hybrid callus. However, 16% of cells in an approximately two-month-old callus still contained two parental types of chloroplast. Nevertheless, we found that only 9.5% of shoots which differentiated from an equivalent callus contained two parental types of chloroplast. These results show that a cell with two kinds of chloroplast has less ability to differentiate into shoots than a normal cell with only one kind of chloroplast. It is reported that in regenerated plants, 95% of parasexual hybrids have only one parental type of chloroplast (Chen et al 1977). This may also be explained by the random chloroplast segregation during the development of a callus and selection of cells with only one type of chloroplast genome to differentiate into shoots.

Acknowledgements

This research was partly supported by a Grant-in-aid for Scientific Research from The Ministry of Education, Science and Culture, Japan.

References

Akada S, Hirai A 1983 Studies on the mode of separation of chloroplast genomes in parasexual hybrid calli. II. Heterogeneous distribution of two kinds of chloroplast genomes in hybrid callus. Plant Sci Lett 32:95–100

Akada S, Hirai A 1986a Studies on the mode of separation of chloroplast genomes in parasexual hybrid calli. III. Random separation of two types of chloroplast genomes in hybrid callus. Jpn J Genet 61:437–445

Akada S, Hirai A 1986b Chloroplast genomes in hybrid calli derived from cell fusion: a novel system to study chloroplast segregation in hybrid calli. Mantell SH et al (eds). The new chondriome: chloroplast and mitochondrial genomes. Longman, Essex, UK, p 290–298

Akada S, Hirai A, Uchimiya H 1983 Studies on mode of separation of chloroplast genomes in parasexual hybrid calli. I. Fraction I protein composition in unseparated hybrid callus. Plant Sci Lett 31:223–230

Carlson PS, Smith HH, Dearing R 1972 Parasexual interspecific plant hybridization. Proc Natl Acad Sci USA 69:2292–2294

Chan P-H, Wildman SG 1972 Chloroplast DNA codes for the primary structure of the large subunit of Fraction I protein. Biochim Biophys Acta 277:677–680

Chen K, Wildman SG, Smith HH 1977 Chloroplast DNA distribution in parasexual hybrid as shown by polypeptide composition of Fraction I protein. Proc Natl Acad Sci USA 74:5109–5112

Glimelius K, Chen K, Bonnett HT 1981 Somatic hybridization in *Nicotiana*; segregation of organellar traits among hybrid and cybrid plants. Planta (Berl) 153:504–510

Ichikawa H, Akada S, Hirai A 1984 Correlation between the type species of chloroplast DNA and that of the large subunit of Fraction I protein in parasexual hybrid calli. Jpn J Genet 59:315–322

Kawashima N, Wildman SG 1972 Studies on Fraction I protein. IV, Mode of inheritance of primary structure in relation to whether chloroplast or nuclear DNA contains code for a chloroplast protein. Biochim Biophys Acta 262:42–49

Medgyesy P, Fejes E, Maliga P 1985 Interspecific chloroplast recombination in a *Nicotiana* somatic hybrid. Proc Natl Acad Sci USA 82:6960–6964

Schiller B, Herrmann RG, Melchers G 1982 Restriction endonuclease analysis of plastid DNA from tomato, potato and some of their somatic hybrids. Mol & Gen Genet 186:453–459

DISCUSSION

Galun: In the Michaelis equation which you showed, the decisive number is the value of n, the initial number of chloroplasts in each cell. Because we really don't know how many there are, we cannot derive from this equation how many divisions are required for the completion of sorting-out. Moreover, not all of the chloroplasts which are there have the ability to divide. So there may be even fewer than five, we really don't know; I wouldn't take fifteen as a number from the literature.

My second remark is that you can now regulate the ratio of the two types of chloroplasts by using mutants which have resistances to antibiotics and then exposing the callus to medium containing those antibiotics. For instance, you use a medium containing both streptomycin and spectinomycin, and you use two types of protoplasts, one of which has chloroplasts that are streptomycin resistant, and the other chloroplasts that are spectinomycin resistant. Then you combine the two protoplasts and in the callus that develops and regenerates, there will be a very high ratio of plants which are heteroplastidic.

Hirai: This is possible but in this case I didn't use any drug resistance. The number of chloroplasts was checked by fluorescence microscopy and the number was not so bad.

Galun: Yes, but not all of those present survive, because isolation of protoplasts damages the ability of the chloroplast to divide.

Hirai: Yes.

Scowcroft: We had this problem some years ago when we were looking at the segregation of *N. debneyii* chloroplasts. The somatic hybrids as identified by isozyme analysis possessed only one or other of the parental chloroplasts, as determined by restriction endonuclease analysis, that is, complete segregation had occurred. Given that the fusing protoplasts contained the high number of chloroplasts characteristic of leaf cells, there were insufficient cell generations from fusion to plant regeneration to enable complete fixation by random drift. The only explanation we could develop was the presence of a bottleneck during culture of fusion products. Such a bottleneck is seen in meristematic regions of developing plants. The number of proplastids per meristematic cell can be as low as ten, cells of photosynthetic leaf tissue can have as many as sixty chloroplasts.

Hirai: But I am talking about cell division from the fused cells to a callus, not meristem.

Galun: Yes, but you select for shoots.

Scowcroft: Once you select for shoots, it is a meristem.

Hirai: From the data of the previous experiment, I assume that cells which have two kinds of chloroplasts have less ability to differentiate. I said that random segregation is only from the fused cells to callus, as shown by the early part of the experiment using the formula.

Scowcroft: Yes, that may be random but the final expectation of what you get is a function of n, where n is the number of chloroplasts.

Hirai: No, the formula in question is only for callus development.

Potrykus: I think it was a beautiful clear study of this question, which has been lacking for a long time. However, being a botanist I would be surprised to find a meristematic cell with chloroplasts, meristematic cells have proplastids which are difficult to count. An additional point is that I would be very happy if I could find clear data on the question of whether differentiated chloroplasts do indeed divide and give rise to the population of chloroplasts in a single cell-derived clone or whether the population of chloroplasts builds up from proplastids. I think there is no clear experimental answer to this, as yet. Counting chloroplasts may be nice but it doesn't help to decide the n number and I think the n-type calculation is based on rather weak grounds. But I think the data which we have seen from Dr Hirai are very good.

Scowcroft: The proplastids in meristematic cells can be quantified by staining their DNA content.

Potrykus: If you stick to the data we have seen, it was a nice demonstration that there is random sorting of plastids to be neutral even in callus, which to me is new information. I think the number of plants containing both types of plastids is surprisingly high compared to the literature as far as I know, because in most cases people found exclusively one or the other type and only in very rare cases was there a mixed population of plastids. This also shows that Dr Hirai is well in control of this experimental system.

Chaleff: Isn't it possible that the shoots are chimaeras and not derived from single cells as you assume? It could be that the 10% of the shoots containing both types of chloroplasts are actually composed of two different types of cells rather than of one type of cell containing two types of chloroplasts.

Hirai: We found several shoots which contain two kinds of chloroplasts but I don't know whether they are chimaeric or not.

Hall: In the first part of your talk you described this very nice technique of going to single protoplasts; why didn't you do that for the second part of your experiment? That would avoid this possibility of having chimaeric plants.

Hirai: That experiment is very difficult to do and we obtained only five calli grown from only a single protoplast; so we couldn't do it in the second part.

Hall: That's a pity, because it is such a nice way of doing it. My second question is, having got those calli from single protoplasts, when you found that the protein patterns were confusing, why didn't you go back and do the restriction patterns since you now have lots of calli?

Hirai: In that case, the callus was grown from single cells. In the first part of the experiment we have to grind the whole callus, therefore there is nothing left to use for another experiment.

Hall: But you can look at restriction patterns from enzyme digests of small amounts of material to verify its heritage.

Hirai: We analysed Fraction I protein at that time so we had no material left to use for restriction pattern anaysis.

Hall: But you had such a nice system and you had the callus that was derived from single cells; I think it would have answered your question.

Hirai: We are trying to repeat that.

Fowler: Could I express a concern about the use and interpretation of electrophoretograms of isoenzymes and proteins in studies of differentiation and development. All living cells and tissues are in a constant state of dynamic activity, with proteins, mRNA species etc., turning over at often quite different rates. When we prepare an electrophoretogram, it is a snapshot of the situation at one moment in time. If the shot had been taken a moment earlier or later, a quite different picture might have been seen, reflecting the dynamic nature of the system, not necessarily a progression of differentiation and development. In addition, all cells and tissues are constantly responding, through either coarse or fine control mechanisms, to their subtly changing environment. Consequently, changes that we see on an electrophoretogram may be no more a reflection of the process of development than of cells adjusting to a shift, no matter how minor, in their environment. Finally, we should not forget that most cell and tissue cultures are highly heterogeneous; major changes seen in enzymes or proteins from one cell population may mask important developmental events in another population where changes in isozyme level are less obvious but of greater significance.

Hirai: That's why I showed the pattern of the restriction enzyme analysis of chloroplast DNA, because if we look at the protein, that is the expressed product of the gene and if there are two kinds of chloroplasts for some reason only one kind of protein may appear. That's why we compared Fraction I protein and the fourth band of *Bam*HI digests in chloroplast DNA. We are sure that if there are two kinds of chloroplast DNA, we get two kinds of Fraction I protein and whenever we get one kind of Fraction I protein then we get one kind of chloroplast DNA.

Fowler: I would like to see a sequence of gels prepared at different times to gain an idea of the dynamic state of the system. It is difficult to make judgements from isolated snapshots.

Potrykus: I think that all biochemical methods have the disadvantage of requiring many cells. It would be worth investing time to studying this question in more detail. It might, however, be worthwhile to look for plastome mutants. I think there's a nice case by Y. Gleba from 1975 where he fused an albino plastome mutant with a wild-type and he got the expected 'sorting out pattern' in the hybrids regenerated from these fusion hybrids. In this case, one could follow, even in single cells, what happens to the mixed population of plastids. I personally would expect to get more information from such a system with visible markers.

Hirai: In fact I am looking for the the mutant.

Takebe: I don't think that my comment will affect your general conclusion from your experiments, but as far as you have tested using Fraction I protein analysis, you can't exclude the possibility that a very small fraction of the chloroplasts comes from other species. Iwai and others (1981) showed that somatic hybrid plants between two *Nicotiana* species apparently contained chloroplasts of either species, as studied by Fraction I protein analysis. However, when they prepared anther cultures from one somatic hybrid plant, they found that a small number of the resultant haploid progeny contained chloroplasts from the other species.

Hirai: As long as we are using Fraction 1 protein as a marker, we cannot exclude the possibility of contamination by a very small number of other chloroplasts.

Takebe: I would not say contamination.

Hirai: We are now beginning to use the chloroplast DNA for this analysis because we can use specific probes to analyse the differences between two chloroplast DNAs from different chloroplasts.

Barz: You mentioned a green callus, i.e. photosynthetically active material, and that it was derived from tobacco, which is a dicot. There are now procedures to establish photosynthetically active cell cultures for a good number of dicots, but as far as I know, it has never been possible to establish photoautotrophic or photomixotrophic cell cultures from monocots, especially not from the Gramineae. Do you know of any work where people have been successful with green cultures from monocots? Do you have any idea why this is so difficult?

Hirai: I have no idea.

Yamada: We are culturing rice and wheat cells in our laboratory and trying to select photoautotrophic cultures from them. We have occasionally been excited when we have found a green cell, but it has always been a shoot primordia.

Galun: That in itself is good.

Hall: You can get very nice green callus of sorghum.

Yamada: How did you get it?

Hall: I just asked my graduate student to do it! Unfortunately, it won't differentiate so we have not been very interested in it.

Barz: Let's be precise in this case. It has been observed many times in various monocots that the calli turned green but if one looked at these green patches under a microscope they were composed of differentiated or semi-differentiated tissue.

Hall: This sorghum callus is solid green.

Barz: Green is not a sufficient indication of a photosynthetically active cell culture; you have to look at this material under the microscope.

Hall: We didn't study it further, because we were interested in finding embryogenic centres in it and there weren't any. We have not looked at it from the point of view that you looked at and may be we should. I certainly could not exclude the fact that there may be some differentiation but we never saw any.

Scowcroft: I've seen banana cultures, which is a monocot, that are green.

Fowler: I recollect a paper some years ago, where it was reported that the fatty acid structure in the thylakoid membrane systems of cultured cells was very different to that in a functional chloroplast in the whole plant. This resulted in a 'structural failure' in the architecture of the thylakoid membranes so that the light trapping system was not operational.

Barz: There are various reasons that one could cite as the possible cause for the difficulties in establishing green cultures from monocots, especially from the cereals. We have been trying this unsuccessfully with maize, wheat and barley, but I have never worked with sorghum or banana.

References

Iwai S, Natata K, Nagao T, Kawashima N, Matsuyama S 1981 Detection of the *Nicotiana rustica* chloroplast genome coding for the large subunit of Fraction I protein in a somatic hybrid in which only the *N.tabacum* chloroplast genome appeared to have been expressed. Planta 152:478–480

Transformation of plant cells

T.C. Hall and R.T. DeRose

Department of Biology, Texas A&M University, College Station, Texas, 77843-3258 USA

Abstract. Techniques for the isolation and characterization of DNA fragments encoding the genetic information of many proteins of agricultural and medical importance have developed rapidly over the past decade. Several vector systems now permit the transfer of specific DNA sequences to plant cells, where they can be efficiently expressed. The tissue-specificity and temporal fidelity of expression of inserted gene sequences in transgenic plants that result from regeneration of transformed cells are remarkable. Inheritance of the acquired sequences is typically stable and Mendelian.

For many dicotyledons the Ti plasmid of *Agrobacterium tumefaciens* is an effective vector; recent reports indicate that it can also mediate transformation of a range of monocotyledons. Injection of DNA into plant meristems and introduction of DNA into protoplasts (facilitated by chemicals such as polyethylene glycol or by electroporation) are alternative approaches that may be especially suitable for monocotyledonous crops.

Rapid advances can be expected in the spectrum of plants modified by recombinant DNA technology and cell regeneration. Molecular processes of gene regulation and cell development are being elucidated. Many practical applications are evident, including the improvement of nutritional value and processing qualities of crop plant products. Commercial opportunities have already been demonstrated by the development of plants that are resistant to herbicides, insects and plant pathogens.

1988 Applications of plant cell and tissue culture. Wiley, Chichester (Ciba Foundation Symposium 137) p 123–143

Many exciting conceptual and practical opportunities derive from technologies developed over the past decade for transformation of plant cells and the ability to obtain fertile progeny from regenerated plants. A plethora of novel ideas and applications exist; room permits only an overview in this article. For additional information, the excellent review by Fraley et al (1986) provides detailed insight into genetic transformation of higher plants; Hall & Thomas (1987) have reviewed the uses of molecular approaches for manipulation of developmental processes in plants, and the National Research Council (1987) report on Agricultural Biotechnology documents a spectrum of practical opportunities arising from gene transfer technology.

Ti plasmid-based gene transfer technology

Recognition during the 1970's that the proliferation of plant cells in 'crown

123

gall' infections was dependent upon plasmids present in certain strains of the soil bacterium *Agrobacterium tumefaciens* led to extensive characterization of these large (nearly 200 kilobase, kb) tumour-inducing (Ti) replicons. Elegant studies in several laboratories showed that only certain regions of these plasmids are essential for conferring the transformed, tumorous phenotype on infected cells (reviewed in Fraley et al 1986). It is now known that the DNA transferred (T-DNA) from Ti plasmids is bounded by 25 base direct repeat sequences. The presence of only one copy of this sequence is sufficient to permit DNA transfer from the plasmid to the chromosome of the plant host when stimulated by proteins synthesized in response to the activation of the *vir* region of the Ti plasmid by elicitors such as acetosyringone (Stachel et al 1985, Bolton et al 1986, Sheikholeslam & Weeks 1987) produced by plants in response to wounding and infection. Continuing studies on the interaction of elicitors, the *vir* region, and T-DNA transfer are likely to be important in broadening the already wide range of plant species that can be transformed by Ti plasmid-mediated gene transfer.

Once the T-DNA region of the Ti plasmid was identified, it seemed very likely that foreign genes inserted within this region would be transferred to the host chromosome along with the T-DNA. This proved to be the case (Chilton et al 1980, Herrera-Estrella et al 1983); unfortunately, the large size of the Ti plasmid made it difficult to work with (shaking the tube containing a wild-type Ti plasmid can shear the DNA, rendering the plasmid inactive), and gene insertion based on homologous recombination (Ruvkun & Ausubel 1981) with wild-type Ti plasmids was cumbersome. Nevertheless, such systems were valuable in confirming that genes could not only be transferred between different plant species, but also expressed as mRNA and protein. An early example of successful expression from a transferred plant gene was the synthesis of the bean seed protein phaseolin in sunflower callus (Murai et al 1983). The concept of binary Ti plasmids (Hoekema et al 1983) permitted great simplification of Ti-based gene transfer, since the gene of interest can be readily inserted into the T-DNA contained within a small plasmid, which is then co-integrated into *A. tumefaciens* containing a relatively large 'helper' Ti plasmid that bears the *vir* region, but little T-DNA. This approach also assisted in eliminating abnormal growth and, frequently, poor fertility observed in transformed plants. This was due to the deletion of genes encoding enzymes catalysing the synthesis of cytokinin and auxin (Zambryski et al 1983) from the helper plasmid. Removal of these genes precluded selection on the basis of hormone-independent growth of the transformed tissue, and absence of genes encoding the synthesis of opines also eliminated screening on the basis of these compounds; however, superior selectable and screenable markers are now available (see below).

The BIN 19 (Bevan 1984) binary vector system and its derivatives are widely used for genetic engineering of dicotyledonous plants, although many facile alternative Ti-based vectors, such as the integrative (pGV3850) vector of Zambryski et al (1983), the SEV system of the Monsanto group (Fraley et al 1985), and Ri (root-inducing) plasmid-based systems (An et al 1985, Vilaine & Casse-Delbart 1987), are available and have advantages in certain instances. In addition to further studies on the interaction of elicitors with the *vir* region, future research is directed to the development of helper plasmids of greater virulence (to obtain higher transformation frequencies) and wider host range than that of the LBA4404 strain currently used ubiquitously in binary vector systems. The recent finding of Cannon and associates (Buchanan-Wollaston et al 1987) that alternative systems of gene transfer between bacteria and plants exist also promises to broaden the species boundary and facility of plasmid-based gene transfer among plants.

Host range limitations and alternative systems for gene transfer

Although many, if not most, dicots are susceptible to infection by *A. tumefaciens*, other Ti plasmids exist that have differing abilities to mediate T-DNA transfer; one important difference being the host range. Among the major classes (typically defined by the opine produced) of Ti plasmid, nopaline and ocotopine variants have been intensively studied. Sequencing of the 24 595 base T-DNA region of the Ti plasmid (pTi 15955) from *A. tumefaciens* (Barker et al 1983) showed it to have three discrete regions, bounded by two imperfect 24 bp repeats; nopaline-type Ti plasmids have a single T-DNA region. Undoubtedly, other regions of Ti plasmids contribute to their host range and virulence. Because inoculation of monocots with Ti (or Ri) plasmids yielded no readily discernable tumour, it was generally thought that they were not susceptible to *A. tumefaciens* infection and, consequently, that Ti-based systems for gene transfer could not be used for this important group of plants. However, closer inspection revealed that asparagus (Hooykaas-Van Slogteren et al 1984, Hernalsteens et al 1985, Bytebier et al 1987), and other plants in the Liliaceae and Amaryllidaceae produce opines and small swellings at *Agrobacterium*-infected wound sites. These results suggest that certain monocots may be susceptible to transformation by Ti-based approaches; perhaps the key to success is the incorporation of appropriate elicitors of the *vir* region into transformation procedures (Schäfer et al 1987).

Economically, maize (*Zea mays*) is an especially attractive target for genetic engineering, but there is no evidence, as yet, that it is directly susceptible to *A. tumefaciens*. However, Grimsley et al (1987) reported that two weeks after inoculation of maize with DNA constructs containing a tandemly repeated dimer of the maize streak virus (MSV) genome within the T-DNA of pTi C58, symptoms typical of MSV were obtained. MSV usually requires the obligate participation of an insect in its transmission to plants, and this 'agroinfection' procedure shows that Ti plasmid sequences can mediate the delivery of DNA to maize. It is not thought that agroinfection results in the integration of this DNA into the host genome, and its value as a technique for stable transformation of plants remains to be proven.

The direct delivery of DNA to protoplasts, by either polyethylene glycol (Krens et al 1982) or electroporation-mediated techniques, has been shown to result in stable transformation of dicots (Paszkowski et al 1984, Potrykus et al 1985) and monocots (Fromm et al 1985, 1986, Ou-Lee et al 1986). Important advances in the regeneration of rice from protoplasts (Yamada et al 1986, Abdullah et al 1987, Kyozuka et al 1987) indicate that, in combination with direct DNA transfer, recombinant DNA-mediated modification of rice will soon be achieved. Further details of protoplast regeneration and direct gene transfer appear elsewhere in this volume. Another method of direct delivery of DNA into monocot species has been 'macroinjection' of young floral tillers (de la Peña et al 1987), however, this method requires screening of massive quantities of seed without the assurance that the resulting plants will be fertile.

Viral vector systems for delivery of DNA to monocots also promise to be effective. French et al (1986) substituted a DNA sequence encoding chloramphenicol acetyltransferase (CAT) for the sequence encoding the viral coat protein into a cDNA copy of brome mosaic virus RNA3. Infection of barley protoplasts with transcripts of this construction resulted in the expression of CAT. A similar approach has been used by Takamatsu et al (1987) to express CAT in tobacco plants infected with transcripts of tobacco mosaic virus RNA that contained the CAT sequence substituted for that of the viral coat protein. As for the agroinfection approach, these viral systems are unlikely to result in stable transformation, but are important because they permit high levels of transient expression of desired genes.

Selectable and screenable traits

Although remarkable for the successes achieved, the frequency of transformation of plant cells with DNA rarely exceeds 1%, and is often in the range of 10^{-4} to 10^{-6}. As a result, effective systems for the selection of transformed cells are essential. Kanamycin resistance (Kan^r), conferred by expression of the neomycinphosphotransferase II (NPT II) gene is currently by far the most commonly used approach. The addition of kanamycin, and other aminoglycosides such as neomycin and G418, to plant tissue culture media typically results in bleaching and death of the cells. Introduction of the NPT II gene can permit plant cells to withstand kanamycin concentrations of up to 1 mg/ml, depending on the strength of expression. A variety of promoters have been used in these experiments, various versions of the 35S promoter of cauliflower mosaic virus (CaMV) typically being the most effective. Although many successes with *Kan*[r] selection have been reported, plant species vary in their susceptibility to the aminoglycosides and even within *Nicotiana* marked differences have been encountered. Genes conferring resistance to other antibiotics, such as hygromycin (van de Elzen et al 1985), have also been used.

Genes conferring resistance to herbicides are also proving to be very suitable for use in selectable marker systems. Examples are: glyphosate resistance (Shah et al 1986, Fillatti et al 1987), resistance to sulphonylureas (Chaleff & Ray 1984), or to imidazolinone (Shaner & Anderson 1985) herbicides (see Chaleff, this volume).

CAT and NPT have been used as screenable traits, especially to provide estimates of the efficiency with which a given gene construction is transcribed (e.g. the comparison of CaMV 35*S* and nopaline synthetase promoters by Sanders et al 1987) and translated by the transgenic plant cell. The value of both of these assays suffers from the presence of varying levels of endogenous enzyme in untransformed plants. Expensive radioactive substrates were needed to assay CAT activity (Gorman et al 1982) but a relatively inexpensive HPLC assay (Young et al 1985) and a rapid, single tube assay (Neumann et al 1987) for the analysis of CAT expression in transgenic tissues have been developed.

Two screenable markers likely to be extensively used in the future because of their extremely sensitive and rapid assays are the luciferase system of Ow et al (1986) and the β-glucuronidase (GUS) system described by Jefferson et al (1987). The luciferase system appears to be especially well suited for dynamic situations; for example, luminescence occurring upon uptake of the luciferin substrate can be followed photographically. Light will be emitted only from those tissues expressing luciferase, the coding region of this gene being placed under the regulation of the promoter being tested. A wide range of substrates can be used for the GUS assay, since β-glucuronidase catalyses cleavage of the β-glycoside linkage of a variety of natural and synthetic glucuronides. Cleavage of the non-fluorescent substrate 4-methyl umbelliferyl glucuronide by this enzyme yields the intensely fluorescent compound, 4-methyl umbelliferone. Accurate spectrofluorimetric detection is possible using extracts from a very few transformed cells, and linear responses over long time periods permit accurate quantitation of the expression levels. It is likely that substrates with even higher fluorescence yields will be developed, and substrates suitable for histochemical detection would be extremely attractive for studying gene expression at the subcellular level.

Currently, the GUS system is exceptionally valuable for rapid analysis of transient expression, e.g. after electroporation of protoplasts. In general, the background activity seen for the luciferase and GUS assays is lower than that encountered for the CAT and NPT systems.

Gene expression obtained in transgenic plants

Early hopes for gene transfer were principally focused on the possibility that detectable levels of expression would be obtained. However, experiments in which a β-phaseolin gene (encoding the major storage protein in the seed of the French bean, *Phaseolus vulgaris*) was transferred to tobacco using a Ti vector system, revealed that not only did the transferred gene give rise to normal levels of mRNA (and protein), but also developmental regulation of its expression, both in terms of tissue and temporal specificity, was maintained (Hall et al 1985, Sengupta-Gopalan et al 1985). Since the coding region of this gene was flanked by only 863 bp of 5' and 1226 bp of 3' sequence, it was evident that these regions contained the major signals for tissue-specificity. The 5' sequence contains repeats of CCAT and TATA motifs associated with transcriptional regulation of eukaryotic genes (Slightom et al 1983), and a study of β-phaseolin gene expression in transgenic plants after alteration or deletion of these (and other) sequences was seen as an exciting approach to studying the function of such *cis*-acting factors. Deletion of sequences upstream of the CCAT box does not allow β-phaseolin expression in transgenic tobacco seed (Klassy & Hall, in preparation), indicating that the CCAT and TATA regions are not by themselves sufficient for expression. Similar results have been obtained by Chen et al (1986) for a homologous (β-conglycinin) gene from soybean transferred to petunia. It is now clear that not only are the 5' sequences (probably also those within and 3' to the coding region) essential for expression of plant and animal genes, but that factors act in *trans* with these regions to co-ordinately regulate expression. Several such regions have been identified for eukaryotic genes, and Jofuku et al (1987) recently reported that a $5'-ATT^A_TAAT-3'$ sequence was involved in the binding of *trans*-acting factors to the 5' region of soybean lectin as identified by DNase 'footprinting'. Many laboratories studying both animal and plant genes are engaged in experiments to define *cis*- and *trans*-acting elements, and it is essential that the results obtained *in vitro* be confirmed by functional expression tests *in vitro* using transgenic organisms.

Light-regulated expression is clearly of great importance to plants, and very informative studies in this area have been achieved through gene cloning and the use of transgenic plants. The *rbcS* gene, encoding the small subunit (SSU) of carboxylase, has received much attention and attachment of its 5' flanking region to screenable markers such as CAT (Herrera-Estrella et al 1984) has shown that it can confer light-regulated expression. Such experiments have confirmed the existence of enhancers (Timko et al 1985) and silencers (Kuhlemeier et al 1987) in the flanking regions of plant genes. Genes encoding the chlorophyll a/b binding protein have also been studied, and their flanking sequences found to confer light-regulated expression (Nagy et al 1987). The sequence and properties of phytochrome are similarly being elucidated through the elegant studies of Quail and his group (Colbert et al 1985, Hershey et al 1985, Vierstra & Quail, 1985).

Organelle-specific expression is of special importance to plants. Although no definitive reports of direct transformation of the chloroplast or mitochondrial genomes have appeared, important progress has been made in understanding features that enable proteins to be transported to these organelles. The mRNAs for both carboxylase SSU and the light-harvesting chlorophyll a/b protein are transcribed from nuclear genes and translated in the cytoplasm. An N-terminal transit peptide (Chua & Schmidt 1979) on the protein products participates in their targeting to the chloroplast. In the case of SSU this sequence has been shown to be necessary and sufficient for import (Mishkind et al 1985, Van den Broeck et al 1985). Karlin-Neumann and Tobin (1986) have noted a common amino acid framework for these chloroplast-targeting peptides, and it will be interesting to learn how broad a range of proteins can be imported through their mediation. Interestingly, Hurt et al (1986) have found that the SSU transit sequence from *Chlamydomonas reinhardtii* was capable of transporting both a yeast cytochrome oxidase subunit (IV) and mouse dihydrofolate reductase into the mitochondrial matrix of yeast.

Targeting of protein to plant mitochondria was demonstrated by Boutry et al (1987). A CAT-encoding sequence was fused behind a sequence encoding 90 amino acids of the N-terminus of the β-subunit of mitochondrial ATP synthase (a nuclear-encoded protein) from *Nicotiana plumbaginifolia*. This was expressed in transgenic tobacco under the control of a 35*S* promoter. As the authors note, this approach may permit expression of novel traits in mitochondria and assist in resolving aspects of mitochondrial biogenesis and cytoplasmic male sterility.

Questions relating to organelle-specificity of expression can be extended to examination of subcellular targeting of both proteins and RNAs. These considerations are of fundamental importance and such knowledge is critical for the success of many practical goals. The seed protein phaseolin is found only in seeds; its accumulation is targeted to protein bodies, which arise from the pre-existing vacuole. During its transit from the initial site of synthesis on rough endoplasmic reticulum, the phaseolin molecule undergoes several modifications. These include the removal of the N-terminal transit sequence and N-glycosylation (Paaren et al 1987). The β-phaseolin sequence contains two N-glycosylation sites and specific changes in the sugar 'decoration' occur as phaseolin traverses the Golgi and enters the protein body (Sturm et al 1987). Many of these events can be studied in detail through the expression of specifically modified gene sequences introduced into transgenic plants.

Studies on subcellular localization also permit evaluation of the stability of both RNA and protein in various tissues and subcellular locations. It is becoming clear that features of both of these molecules confer stability (and instability); for example, phaseolin expressed in sunflower callus (Murai et al 1983) was rapidly degraded. Experiments in which genes encoding zein, the major storage protein of *Zea mays*, were transferred to sunflower (Matzke et al 1984, Goldsbrough et al 1986) indicated that transcription occurs, but no protein has been detected. At present it is not clear if the primary cause for failure is the instability of zein in dicot tissues, or whether important differences exist in regard to gene expression in monocots and dicots. Support for the latter alternative is found in the recent paper by Ellis et al (1987), where no expression of a CAT coding region placed under control of the monocot promoter for maize *Adh*-1 was obtained in transgenic tobacco plants unless a 5' region of the *ocs* gene (or a similar region of the 35*S* gene) was included in the construction.

In addition to the remarkable fidelity of developmental and tissue-specific regulation frequently observed for transgenes, inheritance of expression typically appears to be Mendelian, and the trait seems to be stably integrated into the new genome. Mendelian segregation of phaseolin in tobacco was established by Sengupta-Gopalan et al (1985), and expression was maintained in subsequent generations. Potrykus et al (1985) demonstrated that Kan^r introduced into *N. tabacum* by direct gene transfer was also stably integrated, and exhibited Mendelian inheritance over at least two generations. In virtually all plant transformation experiments, variation in the level of expression of a given gene has been seen in the transgenic progeny, and many comments have been made concerning 'positional effects', i.e. the location of the foreign gene within the host genome. Experience has shown that concerns about the variability of expression were exaggerated, and that most variation is seen with promoters generally regarded to be constitutively expressed. In fact, few promoters appear to be equally expressed throughout development, and even the 35*S* promoter has been found to vary in expression during the cell cycle (Nagata et al 1987, Sanders et al 1987). Clearly, the position of gene insertion is likely to have some effect on expression and further studies on positional effects are definitely needed. Nevertheless, the general reliability of expression levels and stability of introduced traits warrant optimism that recombinant approaches will result in major contributions to future increases in agricultural productivity.

Practical applications for transgenic plants

For practical modification of crops, DNA containing the desired information must be introduced into a cell that can subsequently be regenerated to yield a mature, fertile plant. As has been described, procedures are now well established for the introduction of foreign genes and their expression in dicots. The major remaining barrier is the limited range of dicots that can be regenerated from transformable cells. In monocots, the regeneration barrier appears to be even greater and much research remains to be done before efficient expression of transferred genes in monocots is achieved. Nevertheless, major advances in regeneration of both monocots and dicots have been made and the list of plant species for which transformation has been reported is now extensive.

An important objective for crop modification is the development of herbicide-resistant plants. Resistance to several different herbicides has now been engineered into a number of crop species. An early report was resistance to glyphosate, a broad-spectrum herbicide, in transgenic tobacco containing a modified *aro*A gene from *Salmonella typhimurum* (Comai et al 1983, 1985); a similar approach has been used to obtain tomato plants that are resistant to 0.84 kg active ingredient/ha (Fillatti et al 1987). Glyphosate kills plants by interfering with 5-enolpyruvylshikimate-3-phosphate (EPSP) synthase involved in the shikimic acid pathway; the *aro*A approach overcomes this through the use of a mutant bacterial gene resistant to this block. An alternative method was used by Shah et al (1986), who overcame glyphosate inhibition by overexpressing the plant EPSP gene. Yet another approach is to detoxify the herbicide, as has been done in the case of phosphinothricin by introducing (under 35*S* promoter regulation) the *bar* gene (Thompson et al 1987) from resistant *Streptomyces hygroscopicus* into transgenic tobacco, potato and tomato plants (De Block et al 1987).

Resistance to pathogens is an attractive commercial goal that is being achieved by the development of transgenic plants. Abel et al (1986) reported that tobacco and tomato plants expressing the coat protein gene of tobacco mosaic virus were resistant to subsequent infection by the virus. A similar cross-protection approach yielded resistance to alfalfa mosaic virus in transgenic tobacco (Loesch-Fries et al 1987) and tomato (Tumer et al 1987) plants expressing the cognate coat protein. Certain virus infections of plants can be attenuated by satellite RNAs that are replicated in the presence of the virus but which often show little or no sequence homology to the helper virus genome. This knowledge has formed the basis for an alternative successful approach to virus resistance: the expression of viral satellite RNA in transgenic plants. An initial report by Baulcombe et al (1986) has been extended by Harrison et al (1987), who have shown that transgenic tobacco plants expressing cucumber mosaic virus satellite RNA are resistant to subsequent infection with cucumber mosaic virus or the related tomato aspermy virus. Gerlach et al (1987) have shown that transgenic tobacco plants expressing tobacco ringspot satellite RNA are resistant to this virus. The satellite approach is especially satisfying in that increased levels of satellite transcript are produced in the transgenic plants upon infection, providing a regulated system that does not appear to be saturated by challenge from the pathogen, a potential problem for the cross-protection approach.

Crop damage from insects is a major cause of economic loss, but the use of insecticidal sprays is ecologically undesirable. Several laboratories have been investigating the use of the protein produced by *Bacillus thuringiensis*, which is toxic to a wide range of lepidopteran insects. Vaeck et al (1987) described success in protecting tobacco by the expression of modified versions of the *B. thuringiensis* toxin from the nopaline synthetase promoter in transgenic tobacco. A major difficulty with this work is the apparent instability of the toxin in plant cells, and differences in the codon usage between plants and *B. thuringiensis*. This exemplifies the need to understand characteristics that confer stability on both mRNA and protein expressed in transgenic tissues.

Many opportunities exist for improving the nutritional and processing qualities of crop products, especially seed proteins. Legume seed proteins, such as phaseolin, are typically low in the essential amino acid methionine, and cereal seed proteins, such as the zeins in maize, are low in lysine. Insertion of codons for the deficient amino acids into the coding regions of the seed protein genes should result in improvement of their nutritional value. An additional approach, being tested in the case of phaseolin, is deletion of the asparagine residues that are crucial for covalent attachment of sugar residues, known to decrease digestibility (Liener & Thompson 1980). However, although storage proteins typically have no known enzymic activity, it is likely that constraints on their structure exist, contributing to the overall sequence conservation of these proteins (Hall et al 1983). Such constraints include their regulated hydrolysis during germination, the need to be water-insoluble during storage, and a requirement for stability to environmental conditions (including insects and other pests) during seed dormancy. As a consequence, amino acid insertions and deletions may need to be placed in sites that have no effect on the overall structure of these proteins, again accentuating the need for basic studies in cooperation with applied research.

The list of additional opportunities for development of novel plants through genetic engineering is long and varied. It includes alteration of taste through the addition of proteinaceous sweeteners such as thaumatin (Eden et al 1982) or monellin (Ogata et al 1987), and resistance to stress conditions through the introduction of heat-shock genes (Baumann et al 1987) or genes such as *Adh*-1 that may confer resistance to anaerobic conditions (Ellis et al 1987). The expression of proteins in new locations, e.g. seed proteins in leaves, may be an important agronomic opportunity and, as determinants of morphology become known, modification of plant structure raises many possibilities. Understanding the molecular basis for incompatibility (Collins et al 1984, De Verna et al 1987) and cytoplasmic male sterility (Laughnan & Gabay-Laughnan, 1983, Pring et al 1987) is likely to permit new approaches to heterosis and plant breeding in general. Several genes can be introduced via established transformation processes: these may be added as a multiple gene construct, or as subsequent transformation of an already transformed plant, for example, a plant engineered to be virus resistant may subsequently be engineered to be insect resistant.

In summary, we are just beginning to witness the production of transgenic plants visibly exhibiting desirable traits. In many instances, expectations of the business community have exceeded the ability of scientists to realize promised objectives. Undoubtedly, financial support for genetic engineering of crop plants will be reinforced by the novel plants now available, and considerable increases in research activity in plant biotechnology can confidently be expected in the coming years. However, attainment of commercial goals will undoubtedly require expanded efforts towards understanding basic processes of gene regulation and function in plant growth and development.

Acknowledgements

We thank M. Guiltinan, R. Klassy, J. Anthony and M. Battraw for allowing their data to be presented here. This research was supported by grants from the Texas Advanced Technology Research Program, Rhône-Poulenc Associates and the National Science Foundation (DCB-8602497).

References

Abdullah R, Cocking EC, Thompson JA 1987 Efficient plant regeneration from rice protoplasts through somatic embryogenesis. Bio/Technol 4:1087–1090

Abel PP, Nelson RS, De B et al 1986 Delay of disease development in transgenic plants that express the tobacco mosaic virus coat protein. Science (Wash DC) 232:738–743

An G, Watson BD, Stachel S, Gordon MP, Nester EW 1985 New cloning vehicles for transformation of higher plants. EMBO (Eur Mol Biol Organ) J 4:277–284

Barker RF, Idler KB, Thompson DV, Kemp JD 1983 Nucleotide sequence of the T-DNA region from the *Agrobacterium tumefaciens* octopine Ti plasmid pTi15955. Plant Mol Biol 2:335–350

Baulcombe DC, Saunders GR, Bevan MW, Mayo MA, Harrison BD 1986 Expression of biologically active viral satellite RNA from the nuclear genome of transformed plants. Nature (Lond) 321:446–449

Baumann G, Raschke E, Bevan M, Schöffl F 1987 Functional analysis of sequences required for transcriptional activation of a soybean heat shock gene in transgenic tobacco plants. EMBO (Eur Mol Biol Organ) J 6:1161–1166

Bevan M 1984 Binary *Agrobacterium* vectors for plant transformation. Nucl Acids Res 22:8711–8721

Bolton GW, Nester EW, Gordon MP 1986 Plant phenolic compounds induce expression of the *Agrobacterium tumefaciens* loci needed for virulence. Science (Wash DC) 232:983–985

Boutry M, Nagy F, Poulsen C, Aoyagi K, Chua N-H 1987 Targeting of bacterial chloramphenicol acetyltransferase to mitochondria in transgenic plants. Nature (Lond) 328:340–342

Buchanan-Wollaston V, Passiatore JE, Cannon F 1987 The *mob* and *oriT* mobilization functions of a bacterial plasmid promote its transfer to plants. Nature (Lond) 328:172–175

Bytebier B, Deboeck F, De Greve H, Van Montagu M, Hernalsteens J-P 1987 T-DNA organization in tumor cultures and transgenic plants of the monocotyledon *Asparagus officinalis* Proc Natl Acad Sci USA 84:5345–5349

Chaleff RS, Ray TB 1984 Herbicide-resistant mutants from tobacco cell cultures. Science (Wash DC) 223:1148–1151

Chaleff RS 1988 Herbicide-resistant plants from cultured cells. In: Applications of plant cell and tissue culture. Wiley, Chichester (Ciba Found Symp 137) p 3–20

Chen ZL, Schuler MA, Beachy RN 1986 Functional analysis of regulatory elements in a plant embryo-specific gene. Proc Natl Acad Sci USA 83:8560–8564

Chilton M-D, Saiki RK, Yadav N, Gordon MP, Quetier F 1980 T-DNA from *Agrobacterium* Ti plasmid is in the nuclear DNA fraction of crown gall tumor cells. Proc Natl Acad Sci USA 77:4060–4064

Chua N-H, Schmidt GW 1979 Transport of proteins into mitochondria and chloroplasts. J Cell Biol 81:461–483

Colbert JT, Hershey HP, Quail PH 1985 Phytochrome regulation of phytochrome mRNA abundance. Plant Mol Biol 5:91–101

Collins GB, Taylor NL, De Verna JW 1984 *In vitro* approaches to interspecific hybridization and chromosome manipulation in crop plants. In: Gustafson JP (ed) Proc 16th Stadler Genet Symp Gene Manipulation Plant Improvement Plenum, New York, p 323–383

Comai L, Sen L, Stalker D 1983 An altered *aro*A gene product confers resistance to the herbicide glyphosate. Science (Wash DC) 221:370–371

Comai L, Facciotti D, Hiatt WR, Thompson G, Rose R, Stalker D 1985 Expression in plants of a mutant *aro*A gene from *Salmonella typhimurum* confers resistance to glyphosate. Nature (Lond) 317:741–744

De Block M, Botterman J, Vandewiele M et al 1987 Engineering herbicide resistance in plants by expression of a detoxifying enzyme. EMBO (Eur Mol Biol Organ) J 6:2513–2518

de la Peña A, Lörz H, Schell J 1987 Transgenic rye plants obtained by injecting DNA into young floral tillers. Nature (Lond) 325:274–276

De Verna JW, Myers JR, Collins GB 1987 Bypassing prefertilization barriers to hybridization in *Nicotiana* using *in vitro* pollination and fertilization. Theor Appl Genet 73:665–671

Eden L, Heslinga L, Klok R et al 1982 Cloning of cDNA encoding the sweet-tasting plant protein thaumatin and its expression in *Esherichia coli*. Gene (Amst) 18:1–12

Ellis JG, Llewellyn DJ, Dennis ES, Peacock WJ 1987 Maize *Adh-1* promoter sequences control anaerobic regulation: addition of upstream promoter elements from constitutive genes is necessary for expression in tobacco. EMBO (Eur Mol Biol Organ) J 6:11–16

Fillatti J, Kiser J, Rose R, Comai L 1987 Efficient transfer of a glyphosate tolerance gene into tomato using a binary *Agrobacterium tumefaciens* vector. Bio/Technol 5:726–730

Fraley RT, Rogers SG, Horsch RB et al 1985 The SEV system: a new disarmed Ti plasmid vector system for plant transformation. Bio/Technol 3:629–635

Fraley RT, Rogers SG, Horsch RB 1986 Genetic transformation in higher plants. CRC Crit Rev Plant Sci 4:1–46

French R, Janda M, Ahlquist P 1986 Bacterial gene inserted in an engineered RNA virus: efficient expression in monocotyledonous plant cells. Science (Wash DC) 231:1294–1297

Fromm ME, Taylor LP, Walbot V 1985 Expression of genes transferred into monocot and dicot plant cells by electroporation. Proc Natl Acad Sci USA 82:5824–5828

Fromm ME, Taylor LP, Walbot V 1986 Stable transformation of maize after gene transfer by electroporation. Nature (Lond) 319:791–793

Gerlach WL, Llewellyn D, Haseloff J 1987 Construction of a plant disease resistance gene from the satellite RNA of tobacco ringspot virus. Nature (Lond) 328:802–805

Goldsbrough PB, Gelvin SB, Larkins BA 1986 Expression of maize zein genes in transformed sunflower cells. Mol & Gen Genet 202:374–381

Gorman CM, Moffat L, Howard B 1982 Recombinant genomes which express chloramphenicol acetyltransferase in mammalian cells. Mol Cell Biol 2:1044–1051

Grimsley N, Hohn T, Davies JW, Hohn B 1987 *Agrobacterium*-mediated delivery of infectious maize streak virus into maize plants. Nature (Lond) 325:177–179

Hall TC, Thomas TL 1987 Molecular approaches for the manipulation of developmental processes in plants. In: Atkin RK (ed) Hormone action in plant development: a critical appraisal. Proc 10th Long Ashton Symp, p 287–297

Hall TC, Slightom JL, Ersland DR et al 1983 Phaseolin: nucleotide sequence explains molecular weight and charge heterogeneity of a small multigene family and also assists vector construction for gene expression in alien tissue. In: Ciferri O, Dure L (eds) Structure and function of plant genomes. Plenum, New York, p 123–142

Hall TC, Reichert NA, Sengupta-Gopalan C et al 1985 Regulation of bean β-phaseolin gene expression in yeast and tobacco seed. In: van Vloten-Doting L et al (eds) Organization and expression of the plant genome. Plenum, New York, p 517–529

Harrison BD, Mayo MA, Baulcombe DC 1987 Virus resistance in transgenic plants that express cucumber mosaic virus satellite RNA. Nature (Lond) 328:799–802

Hernalsteens J-P, Thia-Toong L, Schell J, Van Montagu M 1985 An *Agrobacterium*-transformed cell culture from the monocot *Asparagus officinalis*. EMBO (Eur Mol Biol Organ) J 3:3039–3041

Herrera-Estrella L, Depicker A, Van Montagu M, Schell J 1983 Expression of chimaeric genes transferred into plant cells using a Ti-plasmid-derived vector. Nature (Lond) 303:209–213

Herrera-Estrella L, Van den Broeck G, Maenhaut R et al 1984 Light-inducible and chloroplast-associated expression of a chimaeric gene introduced into *Nicotiana tabacum* using a Ti plasmid vector. Nature (Lond) 310:115–120

Hershey HP, Barker RF, Idler KB, Lissemore JL, Quail PH 1985 Analysis of cloned cDNA and genomic sequences for phytochrome: complete amino acid sequences for two gene products expressed in etiolated *Avena*. Nucl Acids Res 13:8543–8559

Hoekema A, Hirsch PR, Hooykaas PJJ, Schilperoort RA 1983 A binary plant vector strategy based on separation of *Vir*- and T-region of the *Agrobacterium tumefaciens* Ti plasmid. Nature (Lond) 303:179–180

Hooykaas-Van Slogteren G, Hooykaas P, Schilperoort R 1984 Expression of Ti plasmid genes in monocotyledonous plants infected with *Agrobacterium tumefaciens*. Nature (Lond) 311:763–764

Hurt EC, Soltanifar N, Goldschmidt-Clermont M, Rochaix J-D, Schatz G 1986 The cleavable pre-sequence of an imported chloroplast protein directs attached polypeptides into yeast mitochondria. EMBO (Eur Mol Biol Organ) J 5:1343–1350

Jefferson RA, Kavanagh TA, Bevan MW 1987 Beta-Glucuronidase as a sensitive and versatile gene fusion marker in higher plants. EMBO (Eur Mol Biol Organ) J 6:3901–3909

Jofuku KD, Okamuro JK, Goldberg RB 1987 Interaction of an embryo DNA binding protein with a soybean lectin gene upstream region. Nature (Lond) 328:734–737

Karlin-Neumann GA, Tobin EM 1986 Transit peptides of nuclear-encoded chloroplast proteins share a common amino acid framework. EMBO (Eur Mol Biol Organ) J 5:9–13

Krens FA, Molendijk L, Wullems GJ, Schilperoort RA 1982 *In vitro* transformation of plant protoplasts with Ti-plasmid DNA. Nature (Lond) 296:72–74

Kuhlemeier C, Fluhr R, Green PJ, Chua N-H 1987 Sequences in the pea *rbcS-3A* gene have homology to constitutive mammalian enhancers but function as negative regulatory elements. Genes and Development 1:247–255

Kyozuka J, Hayashi Y, Shimamoto K 1987 High frequency plant regeneration from rice protoplasts by novel nurse culture methods. Mol & Gen Genet 206:408–413

Laughnan JR, Gabay-Laughnan S 1983 Cytoplasmic male sterility in maize. Annu Rev Genet 17:27–48

Liener IE, Thompson RM 1980 *In vitro* and *in vivo* studies on the digestability of the major storage proteins of the navy bean (*Phaseolus vulgaris*). Qual Plant Plant Foods Hum Nutr 30:13–25

Loesch-Fries LS, Merlo D, Zinnen T et al 1987 Expression of alfalfa mosaic virus RNA 4 in transgenic plants confers virus resistance. EMBO (Eur Mol Biol Organ) J 6:1845–1851

Matzke MA, Susani M, Binns AN, Lewis ED, Rubenstein I, Matzke AJM 1984 Transcription of a zein gene introduced into sunflower using a Ti plasmid vector. EMBO (Eur Mol Biol Organ) J 3:1525–1531

Mishkind ML, Wessler SR, Schmidt GW 1985 Functional determinants in transit sequences: import and partial maturation by vascular plant chloroplasts of ribulose-1,5-bisphosphate carboxylase small subunit of *Chlamydomonas*. J. Cell Biol 100:226–234

Murai N, Sutton DW, Murray MG et al 1983 Phaseolin gene from bean is expressed after transfer to sunflower via tumor-inducing plasmid vectors. Science (Wash DC) 222:476–482

Nagata T, Okada K, Kawazu T, Takebe I 1987 Cauliflower mosaic virus 35 S promoter directs S phase specific expression in plant cells. Mol Gen Genet 207:273–279

Nagy F, Boutry M, Hsu M-Y, Wong M, Chua N-H 1987 The 5'-proximal region of the wheat Cab-1 gene contains a 268-bp enhancer-like sequence for phytochrome response. Eur Mol Biol Organ (EMBO) J 6:2537–2542

National Research Council 1987 In: Moses PB (ed) Agricultural biotechnology. National Academy Press, Washington DC, p 1–205

Neumann JR, Morency CA, Russian KO 1987 A novel rapid assay for chloramphenicol acetyltransferase gene expression. Bio/Techniques 5:444–448

Ogata C, Hatada M, Tomlinson G, Shin W-C, Kim S-H 1987 Crystal structure of the intensely sweet protein monellin. Nature (Lond) 328:739–742

Ou-Lee T-M, Turgeon R, Wu R 1986 Expression of a foreign gene linked to either a plant-virus or a *Drosophila* promoter, after electroporation of protoplasts of rice, wheat, and sorghum. Proc Natl Acad Sci USA 83:6815–6819

Ow DW, Wood KV, De Luca M, de Wet JR, Helinski DR, Howell SH 1986 Transient and stable expression of the firefly luciferase gene in plant cells and transgenic plants. Science (Wash DC) 234:856–859

Paaren HE, Slightom JL, Hall TC, Inglis AS, Blagrove RJ 1987 Purification of a seed glycoprotein: *N*-terminal and deglycosylation analysis of phaseolin. Phytochemistry (Oxf) 26:335–343

Paszkowski J, Shillito RD, Saul M et al 1984 Direct gene transfer to plants. EMBO (Eur Mol Biol Organ) J 3:2717–2722

Potrykus I, Paskowski J, Saul MW, Petruska J, Shillito RD 1985 Molecular and general genetics of a hybrid foreign gene introduced into tobacco by direct gene transfer. Mol & Gen Genet 199:169–177

Pring DR, Lonsdale DM, Gracen VE, Smith AG 1987 Mitochodrial DNA duplication/deletion event and polymorphism of the C group of male sterile maize cytoplasms. Theor Appl Genet 73:646–653

Ruvkun GB, Ausubel FM 1981 A general method for site-directed mutagenesis in prokaryotes. Nature (Lond) 289:85–88

Sanders PR, Winter JA, Barnason AR, Rogers SG, Fraley RT 1987 Comparison of cauliflower mosaic virus 35S and nopaline synthase promoters in transgenic plants. Nucl Acids Res 15:1543–1558

Schäfer W, Görz A, Kahl G 1987 T-DNA integration and expression in a monocot crop plant after induction of *Agrobacterium*. Nature (Lond) 327:529–532

Sengupta-Gopalan C, Reichert NA, Barker RF, Hall TC, Kemp JD 1985 Developmentally regulated expression of the bean β-phaseolin gene in tobacco seed. Proc Natl Acad Sci USA 82:3320–3324

Shah D, Horsch RB, Klee HJ et al 1986 Engineering herbicide tolerance in transgenic plants. Science (Wash DC) 233:478–481

Shaner DL, Anderson PC 1985. In: Zaitlin M et al (eds) Biotechnology in plant science: Relevance to agriculture in the eighties. Academic Press, New York, p 287

Sheikholeslam SN, Weeks DP 1987 Acetosyringone promotes high efficiency transformation of *Arabidopsis thaliana* explants by *Agrobacterium tumefaciens*. Plant Mol Biol 8:291–298

Slightom JL, Sun SM, Hall TC 1983 Complete nucleotide sequence of a French bean storage protein gene: Phaseolin. Proc Natl Acad Sci USA 80:1897–1901

Stachel SE, Messens E, Van Montagu M, Zambryski P 1985 Identification of the signal molecules produced by wounded plant cells that activate T-DNA transfer in *Agrobacterium tumefaciens*. Nature (Lond) 318:624–629

Sturm A, Van Kuik JA, Vliegenthart JFG, Chrispeels MJ 1987 Structure, position, and biosynthesis of the high mannose and the complex oligosaccharide side chain of the bean storage protein phaseolin. J Biol Chem 262: 3392–3403

Takamatsu N, Ishikawa M, Meshi T, Okada Y 1987 Expression of bacterial chloramphenicol acetyltransferase gene in tobacco plants mediated by TMV RNA. EMBO (Eur Mol Biol Organ) J 6:307–311

Thompson CJ, Movva NR, Tizard R et al 1987 Characterization of the herbicide-resistance gene *bar* from *Streptomyces hygroscopicus*. EMBO (Eur Mol Biol Organ) J 6:2519–2523

Timko MP, Kausch AP, Castresana C et al 1985 Light regulation of plant gene expression by an upstream enhancer-like element. Nature (Lond) 318:579–582

Tumer NE, O'Connell KM, Nelson RS et al 1987 Expression of alfalfa mosaic virus coat protein gene confers cross-protection in transgenic tobacco and tomato plants. EMBO (Eur Mol Biol Organ) J 6:1181–1188

Vaeck M, Reynaerts A, Höfte H et al 1987 Transgenic plants protected from insect attack. Nature (Lond) 328:33–37

Van den Broeck G, Timko MP, Kausch AP et al 1985 Targeting of foreign protein to chloroplasts by fusion to the transit peptide from the small subunit of ribose-1,5-biphosphate carboxylase. Nature (Lond) 313:358–363

van de Elzen PJ, Townsend J, Lee KY, Bedbrook JR 1985 A chimaeric hygromycin resistance gene as a selectable marker in plant cells. Plant Mol Biol 5:299–302

Vierstra RD, Quail PH 1985 Spectral characterization and proteolytic mapping of native 120 kilodalton phytochrome from *Curcurbita pepo*. Plant Physiol 77:990–998

Vilaine F, Casse-Delbart F 1987 A new vector derived from *Agrobacterium rhizogenes* plasmids: a micro-Ri plasmid and its use to construct a mini-Ri plasmid. Gene 55:105–114

Yamada Y, Zhi-Qi Y, Ding-Tai T 1986 Plant regeneration from protoplast-derived callus of rice (*Oryza sativa* L.). Plant Cell Rep 5:85–88

Young SL, Jackson A, Puett D, Melner M 1985 Detection of chloramphenicol acetyltransferase activity in transfected cells: a rapid and sensitive HPLC based method. DNA (NY) 4:469–475

Zambryski P, Joos H, Genetello C, Leemans J, Van Montagu M, Schell J 1983 Ti plasmid vector for the introduction of DNA into plants cells without alteration of their normal regeneration capacity. EMBO (Eur Mol Biol Organ) J 2:2143–2150

Zambryski P, Joos H, Genetello C, Leemans J, Van Montagu M, Schell J 1983 Ti plasmid vector for the introduction of DNA into plants cells without alteration of their normal regeneration capacity. EMBO (Eur Mol Biol Organ) J 2:2143–2150

DISCUSSION

van Montagu: Did you say that when you followed the expression of phaseolin in sunflower you really showed that it was controlled by glycosylation or was that a working hypothesis for the reduced expression?

Hall: That is only a working hypothesis. We know that in the sunflower expression it was a glycosylated form. The point in the sunflower is that there are no structures like the vacuoles that are present in the normal cotyledon. The level of expression of the RNA was lower. If you take an extract of sunflower callus, grind it up and mix it with native bean phaseolin, the callus extract will rapidly digest the phaseolin. In those days we had not anticipated that, since we had previously taken well known proteases and had considerable trouble to digest this protein. I think the take home message is that when moving any protein into a cell one has to be aware of the fact that the cell does have control systems, one of them being proteases which will chew up those proteins that are not normally resident. So it was not a question of glycosylation in the sunflower, although the phaseolin may have been more susceptible, if it had not been fully glycosylated.

van Montagu: We have cloned several organ-specific rice genes. We have made chimaeric constructions. At the level of RNA, the promoters from rice that we picked up from a random clone bank function in tobacco plants. We demonstrated that these were true rice promoters, since in the rice we find the corresponding RNA. We didn't look at the expression of the protein but at least at the level of RNA you can find monocot promoters that work normally in a dicot.

Hall: It has been known for a long time (Matzke et al 1984, Goldsbrough et al 1986) that there is transcriptional activity of these genes but the protein has not been found. So the real question is, even when we know there is RNA expression, why have we seen no protein? Is it simply a matter of targeting or of stability or is it the total amount of RNA produced?

van Montagu: I think there is a problem of incorrect splicing—that has been shown for some monocot genes, but I wouldn't dare to generalize. I think there are two classes of splicing sequences in monocots, one of which is recognized by the dicot splicing system and one which is not.

Galun: Did you introduce the gene from the cDNA or a genomic clone?

van Montagu: From the gene.

Hall: It is important that one focuses on the question of the normal gene structure. Certainly, when you splice together promoter regions, signal sequences and coding regions you can probably get expression. There are many studies working in that direction. The question is, when you have a sequence that looks for all intents and purposes like a normal gene, i.e. one that has a normal CAAT box, TATA box and coding region, why is that not expressed? Several people have done those experiments, for example with zein genes which do not have any intervening sequences, but as far as I'm aware, the literature contains no report of a monocot protein being expressed from its normal genomic environment in a transgenic plant.

Scowcroft: A number of people are looking at some of these seed storage proteins, particularly those of legumes, to alter the nucleotide sequence to, for example, increase the level of sulphur-containing amino acids, on the supposition that it probably wouldn't make a great difference to that protein. But with the glycosylation requirements, the possible targeting of such protein to particular parts of the cytoplasm, and their possible use during germination, do you think its likely that one could, for example, enhance sulphur-containing amino acids in phaseolin or any other seed storage protein and maintain it functionally?

Hall: Yes, we were not doing this baculovirus expression experiment simply to look at methionine levels. We are trying to produce enough molecularly pure phaseolin for crystallographic studies. We have worked together with Drs Robert Blagrove and Peter Colman at the CSIRO in Australia and it is possible to crystallize the native protein. We have it to about 5 Å resolution, but that doesn't provide enough definition to generate a good backbone structure. We believe that because the native protein is from a multigene family it is not molecularly pure. So one of the things that we intend doing is to get a crystal structure. This relates to your question: you may be able to put methionine residues into any position or there may be suitable target sites that you can select, for example short repeat sequences present in certain of the molecular forms of phaseolin. However, you may need to know which sites are, for example, on the surface of the protein. I think in several cases, one is going to have to have a better knowledge of the protein structure to make these modifications. I may be somewhat biased, but the seed proteins have a very long history of specialized function and a high degree of sequence conservation, so it may not be feasible simply to put methionine residues in at will, or to remove sugar residues at will. I think that it is going to be necessary in many proteins to understand the real structural chemistry.

Scowcroft: As an extension of that, I know of research groups that are trying to get such seed storage proteins expressed in the leaf of the cell—that probably presents even more difficulties.

Hall: Well if the experiment works that's fine, we don't even need to do all this esoteric work. It is nice to be able to put these proteins behind the cauliflower mosaic virus 35S promoter and the type of experiment that you mentioned has certainly been attractive to several people for some while. For example, to put seed proteins into alfalfa leaves, so that you have enhanced forage, and perhaps even use enhanced seed proteins.

Zenk: If you succeed in creating a phaseolin which is methionine rich, that will create a large sink for methionine metabolism. Do you think that there is an automatic regulatory mechanism within the plant cells which will supply this methionine or would genes involved in methionine synthesis have to be amplified?

Hall: It's a good question; I think that these approaches will give physiologists a field-day looking at source/sink relationships. There has been some work done in pea, for example, on sulphur regimes. When you take such experiments to extremes by using totally sulphur-deficient nutrient sources, then indeed you do see a deficiency in the production of a particular protein. Whether a plant can regulate its use of sulphur and provide more sulphur to the developing seed, we don't know. If one put high sulphur proteins into a legume, and then found that methionine or cysteine was limiting, the next question would be, if you apply sulphur sprays would it now go in? I think that these are further opportunities for research being opened through genetic engineering and tissue culture, which we really have not been able to answer in any rigorous way previously.

Barz: You described to us the elegant way of vaccinating a plant against virus infection, by having the viral protein expressed in the plant which gives some degree of protection. Is anything known about the mechanism of resistance in this case?

Hall: Not to my knowledge. I think we will get better insight into the methods of viral infection of plants through such experiments. Roger Beachy does know that this approach does not protect against RNA inoculations (Abel et al 1986). It has been postulated that perhaps a large amount of coat protein prevents uncoating of the normal variant when it approaches the surface of the leaf.

van Montagu: We already mentioned the considerable interest in new herbicides that could replace the more toxic ones. There are a series from Monsanto, from DuPont, American Cyanamid and Hoechst, that inhibit target enzymes for amino acid biosynthesis. These are total herbicides, this means that they do not distinguish between weeds and crops. If you want to have resistant plants, you have to engineer the resistance into the plants, then these herbicides can be used specifically. Several approaches for engineering this resistance into plants have already been discussed: one can overexpress the target enzymes, or induce a mutation that makes the target enzyme resistant. The approach that we used was inactivation of the herbicide. If we can engineer plants which inactivate the herbicide, we will have constructs that are very efficient. We succeeded, by following this approach, in developing resistance to Bialaphos and BASTA. Bialaphos is a herbicide manufactured through fermentation by the Japanese company Meiji Seika. It is obtained from a *Streptomyces hygroscopicus* strain and the active compound is phosphinothricin. It acts as an analogue of glutamic acid. The German company, Hoechst, synthesizes chemically the herbicide phosphinotricin. Both the compounds are active as bacteriocides, fungicides and herbicides. We looked, together with Charles Thompson and Julian Davies who at that time worked for Biogen, to discover why and how this *S. hygroscopicus* bacterium can survive Bialaphos production. We demonstrated that this bacterium synthesizes an enzyme that inactivates the compound by acetylating the phosphinotricin moiety of the herbicide. We then cloned the corresponding gene (Thompson et al 1987). It turns out that the enzyme is an acetyl transferase, which uses acetyl Coenzyme A as a cofactor. The cloned gene was transferred into plant cells and expressed. Using tobacco plants as a standard model, you see that the transformed plant grows as happily as the control plants. If both of them are sprayed with 5–10 times the amount of herbicide that is used in the field, the control plant is completely killed by ammonia poisoning and the transformed plant grows still perfectly well. I think that the fact that the plant is so rapidly killed by ammonia poisoning (not a living cell left after ten days) is a particular advantage for the use of this herbicide, since there will be less chance of natural resistance arising in weeds. Any plant cell which mutates towards resistance should be killed by the surrounding ammonia-producing cells. But that remains to be proven. We obtained the same high level of herbicide resistance with tomato plants and with potatoes. For tobacco the field trials were done by breeders of the National Tobacco Institute in Bergerac, France. After spraying the fields, the plants that are resistant continued to grow as normal. The non-engineered plants were very sensitive and became shrunken within 10 days. We feel that this construct will have applications in a variety of crops.

Hall: Can you tell me what tolerance level your resistant plants had, the number of times the normal application rate of the herbicide?

van Montagu: We use concentrations that are 5–10 times the commercial application. If we try to go higher, the plants are killed by the detergents in the herbicide mixture. This enzyme is rather stable in plants, so we don't need a high level of expression. Ironically, we have transformations where 1% of the plant protein is this acetyl transferase.

Riley: How many insertions of the transforming gene?

van Montagu: One insert is sufficient.

Riley: And you are allowed to put that in the field without any regulation?

van Montagu: Well that's the advantage of Europe, administration is more sensible. We applied for permission for field trials in Belgium and in France, there was no major problem. We have learned that permission for field trials has been obtained by other laboratories in the U.K.. But even in the U.S., Rohm and Haas obtained permission for release of 'insecticide producing' plants, as did several agrochemical companies, although there they had to comply with absurd working conditions.

Scowcroft: What promoter are you using?

van Montagu: Several have been used. Even a weak promoter such as those of some opine synthesizing genes are sufficient.

Zenk: When you described your strategy I think you missed one really important point, namely that this is degradation of the herbicide. In the future, public pressure will certainly influence us to move in the direction of trying to degrade all organic chemicals that are applied as herbicides. However, in your case the acetylated amino acid which is induced remains in the plant in that form.

van Montagu: It is degraded by soil bacteria, I don't know by which pathways.

Zenk: But it occurs as a residue in the plant

van Montagu: Hoechst have determined the toxicity of this compound as a residue in the plant and it is even lower than that of phosphinotricin itself, which is considered to be non-toxic.

Zenk: Scientifically, I think we all agree, but it is difficult to convince the general public.

van Montagu: I'm afraid the public will have to realize what compounds have been used for the past twenty or more years and are still used at present. Many of these compounds that have been acceptable until now are quite harmful; this is not the case with the new ones being developed.

Barz: You yourself indicated that the toxicology of this compound is not yet fully established.

van Montagu: More tests need to be done, I agree.

Barz: What are the biological properties of the material? It is an acetyl derivative, and is most probably stored in the vacuole, so it's still in the plant cell. This leads me to the question—do you think it is possible to construct a plant which possesses an excretion mechanism for the modified metabolite?

van Montagu: Our strategy is that Hoechst or Meiji should do this!

Ohyama: This enzyme was originally not present in the plant cell, so why does the plant cell express it?

van Montagu: Frequently when new genes are introduced into a plant cell, new enzymes are made which were not present in the cell before. In this case, a bacterial gene was introduced under the control of a constitutive promoter.

Ohyama: But this gene is not essential for plant development.

van Montagu: No, I agree that if a given enzyme is not needed, why would plants after a certain time not stop making it. It can only be done experimentally, using an engineered constitutive promoter. We have plants in which nopaline synthesis has been induced. We have had these plants since 1979, I don't know how many generations they have been through, but nopaline synthesis continues. People in Gatersleben, East Germany, have analysed neomycin phosphotranferase activity in progeny of plants engineered for antibiotic resistance (Müller et al 1987). In a given tobacco plant, all 45000 seeds obtained were tested for this enzyme. On average, there are one or two seeds where the kanamycin resistance has been lost, possibly through methylation of the gene. So loss of expression can happen but it is a very rare event.

Fowler: Marc's results are very interesting compared with animal cell systems where there are examples of constitutive expression which after a short time destabilizes. The metabolism of the animal cell appears to be markedly affected by the new gene system incorporated. These results are therefore very important for the prospect of continued expression in plant systems.

References

Abel PP, Nelson RS, De B et al 1986 Delay of disease development in transgenic plants that express the tobacco mosaic-virus coat protein gene. Science (Wash DC) 232:738–743

Goldsbrough PB, Gelvin SB, Larkins BA 1986 Expression of maize zein genes in transformed sunflower cells. Mol & Gen Genet 202:374–381

Legrand M, Kauffman S, Geoffroy P, Fritig B 1987 Biological function of 'pathogenesis-related' proteins: four tobacco PR-proteins are chitinases. Proc Natl Acad Sci USA 84:6750–6754

Matzke MA, Susani M, Binns AN, Lewis ED, Rubenstein I, Matzke AJM 1984 Transcription of a zein gene introduced into sunflower using a Ti-plasmid vector. EMBO (Eur Mol Biol Organ) J 3:1525–1531

Müller AJ, Mendel RR, Schiemann J, Simoens C, Inze D 1987 High meiotic stability of a foreign gene introduced into tobacco by Agrobacterium-mediated transformation. Mol & Gen Genet 207:171–175

Thompson CJ, Movva NR, Tizard R et al 1987 Characterization of the herbicide-resistance gene bar from Streptomyces hygroscopicus. EMBO (Eur Mol Biol Organ) J 6:2519–2523

Direct gene transfer to plants

Ingo Potrykus

Institute for Plant Sciences, Swiss Federal Institute of Technology, ETH Zentrum LFV-E20, CH-8092 Zürich, Switzerland

Abstract. Gene transfer by biological vectors is limited by the restricted host range of a given vector. We have developed and exploited a vector-independent method for gene transfer. Incubation of protoplasts with genes under the control of plant gene expression signals leads to high frequencies of stable integrative transformation. The foreign gene is transmitted to sexual offspring and is inherited according to Mendelian laws. The integration of the foreign gene occurs at random sites; in more than 77% of primary transgenic plants studied this was at one locus, but integration at two, three and more independent or linked loci has also been found, as well as cases indicating maternal inheritance. In the majority of transgenic plants analysed, the foreign gene is absolutely stable; so far over eight sexual generations and for nearly three years without any selective pressure. However, transgenic plants expressing various degrees of instability or loss of the foreign gene have also been found. Transformation frequency can reach values of up to 10% of non-selected protoplast-derived clones, independent of which of two optimized methods is used. One is a combination of electroporation and polyethylene glycol treatment, the other a sequential treatment with magnesium ions and polyethylene glycol. The transformation frequency is, however, also species and genotype dependent: the high levels achieved with *Nicotiana tabacum* SR1 were never reached with *N. plumbaginifolia* or *Petunia hybrida*. Direct gene transfer is, apparently, possible with protoplasts from any plant species. Treatment with mixtures of selectable and non-selectable genes led to co-transformation rates of up to 88%. Treatment of protoplasts with sheared or partially digested total genomic DNA from a plant carrying one copy of a dominant, selectable marker gene led to the transfer, integration and expression of this gene. *In situ* hybridization of radioactively labelled probes of the foreign gene to metaphase chromosomes could be used to visualize the location of the gene. 5′ and 3′ deletions of a selectable gene with overlapping stretches of homology have been used to study homologous recombination within plant cells. Stable integration of non-functional 5′ deletions of the same gene into the host genome and subsequent transformation with complementing 3′ deletions were used to demonstrate gene targeting in plants. Microinjection of a marker gene into microspore-derived proembryos produced transgenic plants in *Brassica napus*.

1988 Applications of plant cell and tissue culture. Wiley, Chichester (Ciba Foundation Symposium 137) p 144–162

We are interested in studying the organization and regulation of the plant genome. We would like to be able to alter plant genomes in a predictable way, especially those of the economically important cereal crop species. To do so, we require a routine and efficient gene transfer protocol, which is applicable to cereals. Experience suggested that the available biological vector systems, such as bacteria and viruses, might not be appropriate. We therefore aimed at establishing a generally applicable gene transfer protocol based on the use of protoplasts. Descriptions of the first transgenic cereals produced by direct gene transfer (not from our laboratory) will soon appear in the literature. The developments of the last year (1987) have shown that direct gene transfer can offer interesting advantages (e.g. gene targeting, co-transformation) even in non-cereal transformation experiments. In our opinion, the real breakthrough in gene transfer in cereal plants will, however, come from microinjection of genes into cells with morphogenetic potential, e.g. microspore-derived embryos (Neuhaus et al 1987).

The basic experiment

This is described in detail in Paszkowski et al (1984). Leaf protoplasts of tobacco were incubated with a hybrid selectable marker gene conferring resistance to kanamycin and treated with polyethylene glycol (PEG) to facilitate uptake of the DNA. Subsequent culture under lethal concentrations of kanamycin led to the recovery of resistant protoplast-derived clones, which could be induced to regenerate (on kanamycin) shoots and complete plants. Southern blot analysis and assay for neomycin phosphotransferase activity proved that the novel phenotype was indeed the consequence of uptake, integration and expression of the foreign gene. Analysis of sexual offspring revealed that the foreign gene was maintained and transmitted through meiosis. The transformation frequency was one in one million protoplasts treated.

Protocols for routine high transformation frequencies

These are described in detail in Shillito et al (1985), Negrutiu et al (1987) and Krüger-Lebus & Potrykus (1987). The protocol of Shillito et al combines electroporation with PEG and heat shock treatment: protoplasts are suspended in 0.4 M mannitol containing 6mM $MgCl_2$ and 0.1% MES (2-[N-morpholino] ethanesulphonic acid) buffer pH 5.6 at a population density of about 10^6/ml. $MgCl_2$ is added to bring the electrical resistance into the range of 1200 ohms. Following heat shock (five minutes at 45 °C), the gene is added to a final concentration of 10µg/ml and carrier DNA to 40 µg/ml, followed by PEG (M_r 6000) to a final concentration of 8%. Electroporation is applied with three high voltage pulses at 10 second intervals, which gives 50% survival. The protocol of Negrutiu et al combines a magnesium pre-treatment with PEG treatment at a high concentration: Protoplasts are washed in a Na-Ca-K-Cl solution (W5), transferred into a mannitol solution containing 15 mM $MgCl_2$, heat shocked, followed by addition of the DNA and PEG (M_r 4000) to a final concentration of 20%. In the protocol of Krüger-Lebus, the protoplasts are suspended in osmoticum (mannitol and 50 mM $CaCl_2$), the DNA and carrier DNA are added and incubated for one minute at room temperature. Then PEG 6000 is added to a final concentration of 20%, the protoplasts incubated for 10 minutes at room temperature and transferred to culture medium.

Transformation frequency

We relate the transformation frequency to the number of possible clones. The calculation is based on either screening without prior selection of a few hundred randomly picked clones, or a comparison of the number of clones surviving selection with the number in an unselected aliquot of the treatment. With *Nicotiana tabacum* var.Petit Havana SR1, we routinely achieve stable integrative transformation of a few per cent, up to a maximum of 10%. With *Hyoscyamus muticus*, best values were 1%, with *Petunia hybrida* 0.5%, with *N. plumbaginifolia* 0.2% and with *Lolium multiflorum* about 10%. From extensive comparative studies, in particular with *N. tabacum* and *N. plumbaginifolia*, and careful optimization with *Nicotiana* (Negrutiu et al 1987), we have the impression that plant species (and possibly cell types) differ in their competence for integrative transformation.

Host range

There is probably no species-specific barrier to the straightforward physical uptake of naked DNA into protoplasts and its integration into the host genome. It was no problem to transfer genes into graminaceous protoplasts, such as for *L. multiflorum* (Potrykus et al 1985a), *Triticum monococcum* (Lörz et al 1985), *Zea mays* (Fromm et al 1985) or *Oryza sativa* (Uchimiya et al 1986).

Integration into the host genome

The foreign genes integrate, obviously via illegitimate recombination and therefore at random sites, into chromosomal DNA. So far, there is no clear case known of integration into either chloroplast DNA or mitochondrial DNA. In 77% of 220 primary transformants studied, integration was at one single locus. The number of integrated copies can range from one to 30 copies or more. Also, in the case of multiple copy integrations, these occur preferentially at one locus; there is, however, also simultaneous integration at independent loci. The integration pattern can be very simple or relatively complex. There is again an effect of the plant species and the cell type on the integration pattern.

Localization of the foreign gene in the host genome

In plants, localization of the integrated foreign gene via linkage analysis using marker genes is difficult. We have therefore developed an alternative method of *in situ* hybridization of radioactively labelled probes to metaphase chromosomes (Mouras et al 1987). Protoplasts from root meristems are air dried on slides and co-denatured with a probe of the gene under study at high specific activity. To get a clear signal requires long exposure, therefore a very clean background is essential.

Expression of the integrated gene

The level of expression is influenced by the expression signal linked to the foreign gene, and by other parameters. From work with *Agrobacterium tumefaciens*-based gene transfer, a wealth of knowledge on regulated promoter sequences is available. Our own experience is limited to the expression signals from the 19S and 35S transcripts of cauliflower mosaic virus. From the observation that transgenic clones carrying one copy of the foreign gene may have far higher levels of expression than others with 30 copies, or vice versa, it is obvious that so-called position effects play a role too.

Inheritance of the integrated gene

The bacterial gene for neomycin phosphotransferase, when under the control of the constitutive viral 19S and 35S promoters, exhibits a very simple, clear, dominant phenotype on kanamycin: seedlings (and plants) containing the gene stay green, those lacking the gene bleach completely to white. With seedlings this phenotype is very clear within seven days. The large number of seeds produced per flower in *N. tabacum* (up to 3000) in the majority of the transgenic plants (direct gene transfer has no fertility penalty), in combination with the simple phenotype of the gene, facilitated a detailed genetic analysis. In the great majority of transgenic plants analysed (more than 220) the inheritance followed Mendelian rules, and in all cases the inheritance of the phenotype was strictly correlated with the transmission of the foreign DNA as well as with the neomycin phosphotransferase activity (Potrykus et al 1985b). In 77% of the plants, the gene was inherited as a single dominant Mendelian factor. The remaining fraction of the population contained two, three, or more, linked or unlinked genes. Of special interest is a group of plants which was characterized by clear reciprocal differences and maternal inheritance in reciprocal crosses with the untransformed wild type. Molecular proof for extra-chromosomal inheritance is, however, not available.

Stability of the foreign gene

The fate of the foreign gene has been studied, so far, through seven sexual generations and for more than 30 months in the absence of selective pressure. The predominant effect is of stable inheritance. In stable clones we did not discover any sign of somatic or meiotic inactivation or loss of the functional gene; this is true even for the non-functional copies integrated together with the functional ones.

Instability of the foreign gene

Besides clones with a high degree of stability there were others which showed considerable instability, as exemplified by the following example: a self cross of a homozygotic offspring of a hemizygous primary transformant produced four sensitive seedlings amongst 426 tested. Sixty further R_2 populations of the same R_1 plant, produced by selfing 60 flowers from primary, secondary and tertiary inflorescences all produced seedling populations containing approximately 0.5% sensitive individuals. One resistant R_2 plant tested further produced R_3 populations again with 0.5% sensitive seedlings. Three resistant R_3 plants were used to generate R_4 populations. Two of them continued to produce 0.5% sensitive offspring; one, however, produced about 10% sensitive progeny. The next generation from resistant plants of this population yielded an entire spectrum of plants, from totally resistant, through various ratios of resistant to sensitive, to only sensitive. The molecular basis for this instability has not yet been studied. We have, however, found examples where the band representing the functional copy of the gene has been lost, whilst the non-functional copies have been maintained. We have also found cases where the functional band has been maintained and the non-functional ones have been lost.

Co-transformation

This is described in detail in Schocher et al (1986). Recovery of transgenic clones containing a selectable marker gene is no longer a problem. However, what about those cases where a gene should be transferred, which has no selectable phenotype or which does not even produce a visible (and therefore screenable) phenotype? The high transformation frequency of a few per cent available in *N. tabacum* makes it feasible to screen a small population of approximately 50 non-selected clones for the presence of any gene, and we have done so successfully. This approach is, however, laborious with plants that have lower transformation frequencies. We therefore tested whether the phenomenon of co-transformation would also function in our system. To study this question, a plasmid carrying a zein gene (encoding the maize storage protein) was mixed with the plasmid carrying the kanamycin resistance gene in ratios of 1:1 and 3:1. Kanamycin-resistant clones were recovered and randomly picked clones were analysed by Southern blot analysis for simultaneous presence of the zein DNA. Seventy-two percent of the clones from the 1:1 mixture and 88% from the 3:1 mixture contained DNA hybridizing to the zein probe; there was no hybridization to the control clones. In co-transformation experiments with the nopaline synthase gene, up to 47% of the resistant clones contained and expressed the functional gene. This method of co-transformation is now in routine use in our laboratory for transfer of non-selectable genes.

Gene transfer from total genomes

Reproducible transformations in reconstruction experiments (where we mixed a selectable marker gene with wild-type tobacco DNA) using ratios as low as 10 copies of the marker gene per tobacco genome prompted us to attempt the transfer of a single copy gene from a tobacco genome. To this end, total DNA was isolated from tobacco plants containing one copy of the kanamycin resistance gene stably integrated into the genome. This DNA was sheared or partially digested and added to wild-type tobacco protoplasts. Transgenic plants with this gene were recovered at the expected low frequency of approximately one per three million protoplast-derived clones (T. Müller et al, personal communication). These results show that it should be possible to transfer plant genes from one species to another without prior isolation of the gene. However, the low frequency of transformation will make this feasible only for genes producing a selectable phenotype.

Homologous recombination

Gene transfer, as described so far, was by non-homologous recombination. The frequencies obtained show that this is a very effective process in plants. Non-homologous recombination will always result in random integration. We would like to be able to direct the integration of genes to precise sites in the genome; this would require homologous recombination. To study this question, we constructed 5' and 3' deletion mutants of a selectable marker gene with overlapping stretches of homology of different lengths. If homologous recombination occurred in somatic cells, it should be possible to feed non-functional, complementing 5' and 3' deletion fragments to protoplasts in a co-transformation experiment and recover transgenic clones containing the functional gene. The gene would have been restored by homologous recombination within the overlapping region. Kanamycin-resistant clones were recovered and the frequency depended on the length of homology provided for recombination. This system could be used to study and optimize conditions for homologous recombination (M. Baur et al, personal communication).

Gene targeting in plants

With the experience gained from the experiments on homologous recombination we felt confident that gene targeting — integration of a stretch of foreign DNA at a precise, pre-determined site in the genome — should be possible too. To test this hypothesis, we used the same basic experimental idea as before. In this experiment, however, it was necessary to integrate stably the non-functional 5' or 3' deletions of the selectable marker gene into the genome of the receptor plant, then to treat protoplasts isolated from those receptor plants with complementing 3' or 5' deletions, again selecting thereafter for the reconstitution of the functional gene. The recovery of kanamycin-resistant clones demonstrated that gene targeting is possible in plants. Internal molecular markers were used, which allowed us to discriminate unequivocally against possible misinterpretations of the data. Gene targeting may now be used, not only for the precise delivery of foreign genes to pre-determined sites in the genome, but also for directed mutagenesis and gene tagging (J. Paszkowski et al, personal communication).

Microinjection into microspore-derived embryos

Despite the successes of many techniques of gene transfer, for example, the use of *Agrobacterium tumefaciens* and *Agrobacterium rhizogenes*, 'direct gene transfer', agroinfection, and viral vectors, there are still important groups of plants (e.g. the cereals) where gene transfer is still a problem. A recent experiment of Neuhaus et al (1987) probably offers a solution: microspore-derived embryos of *Brassica napus* at an early state of development (12–16 cells) have been used for gene transfer via microinjection into the nuclei of the majority of the cells of the proembryo. Embryos were regenerated and dot blot analysis and enzyme assay showed that they had been transformed. As had to be expected, these plantlets were transgenic chimaeras. In *Brassica napus* it is possible to resolve the chimaeric situation by producing secondary embryos. This was done and completely transgenic plants were recovered this way. We believe that this will be the preferred method for the production of transgenic cereal plants, reproducibly and under controlled conditions, and experiments along these lines are in progress.

Acknowledgements

The work on direct gene transfer was carried out at the Friedrich Miescher Institute, Basel.

References

Fromm M, Taylor LP, Walbot V 1985 Expression of genes transferred into monocot and dicot plant cells by electroporation. Proc Natl Acad Sci USA 82:5824–5828
Krüger-Lebus S, Potrykus I 1987 A simple and efficient method for direct gene transfer to Petunia hybrida without electroporation. Plant Mol Biol Rep, in press
Lörz H, Baker B, Schell J 1985 Gene transfer to cereal cells mediated by protoplast transformation. Mol & Gen Genet 199:178–182
Mouras A, Saul MW, Essad S, Potrykus I 1987 Localisation by in situ hybridisation of a low copy chimeric resistance gene introduced into plants by direct gene transfer. Mol & Gen Genet 207:204–209
Negrutiu I, Shillito R, Potrykus I, Biasini G, Sala F 1987 Hybrid genes in the analysis of transformation conditions. I. Setting up a simple method for direct gene transfer in plant protoplasts. Plant Mol Biol 8:363–373
Neuhaus G, Spangenberg G, Mittelsten-Scheid O, Schweiger HG 1987 Transgenic rapeseed plants obtained by the microinjection of DNA into microspore-derived embryoids. Theor Appl Genet 74:30–36
Paszkowski J, Shillito RD, Saul MW, Mandak V, Hohn T, Hohn B, Potrykus I 1984 Direct gene transfer to plants. EMBO (Eur Mol Biol Organ) J 3:2717–2722
Potrykus I, Saul MW, Petruska J, Paszkowski J, Shillito RD 1985a Direct gene transfer to cells of a graminaceous monocot. Mol & Gen Genet 199:183–188
Potrykus I, Paszkowski J, Saul MW, Petruska J, Shillito RD 1985b Molecular and general genetics of a hybrid foreign gene introduced into tobacco by direct gene transfer. Mol & Gen Genet 199:169–177
Schocher RJ, Shillito RD, Saul MW, Paszkowski J, Potrykus I 1986 Co-transformation of unlinked foreign genes into plants by direct gene transfer. Bio/Technol 4:1093–1096
Shillito RD, Saul MW, Paszkowski J, Müller M, Potrykus I 1985 High efficiency direct gene transfer to plants. Bio/Technol 3:1099–1103
Uchimiya H, Fushimi T, Hashimoto H, Harada H, Syono K, Sugawara Y 1986 Expression of a foreign gene in callus derived from DNA-treated protoplasts of rice (Oryza sativa). Mol & Gen Genet 204:204–207

DISCUSSION

van Montagu: I appreciate all the work that you have done on the analysis of the inheritance and the demonstration that the inserts behave as single Mendelian loci, but at a given locus, can you tell on average how many copies there are? I know that it is difficult to do this from Southern blots, but you should be able to estimate this from your co-transformation data.

Potrykus: We have everything between one and more than fifty copies. It varies between plant species, there is also an effect of the protoplast type we use. We can partially influence the number of copies integrated by altering the concentration of foreign DNA. But in principle it is not predictable how many copies are integrated.

Scowcroft: I am reticent to suggest that somaclonal variation may be a trivial thing, but possibly that might account for some of the instability you see in a transgenic plant. Simple aneuploidy will lead to instability, either directly or through univalent shift.

Potrykus: But this instability is in a clear diploid case.

Scowcroft: You can also get univalent shift as a result of translocations. Translocations will give an appearance of being diploid but they can cause univalent shift in subsequent generations.

Potrykus: We see a number of possible causes of this instability but we have no data to decide which of the causes is responsible.

Scowcroft: I would also like to ask about the targeting of the genes. Is it possible with your system (with the selectable marker) to make a chimaeric gene including a highly repeated sequence, for example spacer RNA, and to see if you can target it repeatedly to a particular part of the genome, as defined by, say, a restriction fragment length polymorphism.

Potrykus: Certainly we can attempt that once we have optimized the system to get higher targeting frequency. So far the targeting frequency is about 3 x 10^{-3} of the normal transformation frequency. So we have to improve this frequency before we do this kind of experiment. I indicated that this targeting will permit us to study position effects because we can repeatedly put the same or different genes at precisely the same position in the genome.

Riley: I am very encouraged and it looks very optimistic that transformation can take place by the DNA injection processes you describe. One of the things which seemed to worry you was that the transformed material initially would be chimaeric. Must that not always be a possibility with any transformation? The occurrence of chimaerism would depend on the phase of the cell cycle at which transformation occurs. If it occurs in the S phase, then for any chromosome one can understand that the whole chromosome will be transformed. If it occurs in the G_2 phase, then only one of a pair of chromatids will be transformed. What is the actual experience of people who study these phenomena?

Potrykus: We have often found no deviations from normal Mendelian ratios when self-pollinating or backcrossing primary transformants with the wild-type. In 76.6% of our primary transformants analysed, the foreign gene acts like a hemizygous locus.

Riley: So when transformation occurs from a protoplast a heterozygote is created?

Potrykus: Yes, in this respect I would not agree that this is a chimaera. We have studied the production of chimaeras carefully with plants from early development to flowering, and we have not found a single case where we saw any indication of chimaerism.

Riley: That implies that there is no transformation at G_2, otherwise there would be a chimaera, so transformation must always occur either at G_1 or at the S phase.

Potrykus: I don't know whether you can draw this conclusion because you need one copy in one chromatid to get the expression of this gene. If you don't have somatic rearrangements, you will not produce a chimaera because every cell produced will have a stably integrated copy.

Riley: But if only one chromatid of the chromosome were transformed, there would be a chimaera.

Potrykus: Then you would see it in the sexual offspring.

Fowler: Is there a specific period in the cell cycle when transformation or incorporation takes place? Is the rest of the cycle blocked against DNA integration?

Potrykus: We don't see an effect of the cell cycle on transformation. We have the impression that the majority of the mesophyll cells of the tobacco leaf are in G_1. We use arrested mesophyll cells which take at least three days to start the next cell cycle, therefore foreign DNA would have to survive three days before the cell cycle starts, which would cause lots of problems, so we could envisage integration into interphase nuclei.

Galun: The synthesis of DNA in *Nicotiana* protoplasts usually starts about 48 hours after plating.

Potrykus: 48 hours is a long time for a naked piece of DNA to survive in a foreign environment.

Galun: Do you transform it on the day of isolation or do you wait?

Potrykus: We transform freshly isolated protoplasts.

Galun: Have you tried transforming two-day-old protoplasts as well?

Potrykus: We can do it with protoplasts up to seven days old, if we prevent cell wall synthesis. The high transformation frequency is with freshly isolated protoplasts.

Galun: Does the frequency fall after one day?

Potrykus: We add the DNA immediately after isolation, and then we observe the high frequency of transformation. What happens between adding DNA and the observed integration is a black box.

Fowler: If cell wall synthesis is going on, as happens with protoplasts, will that affect transformation during that period of time?

Potrykus: Yes, cell wall synthesis will interfere with transformation, if you allow the cells to progress that far.

Chaleff: If I may return to Sir Riley's question: it would seem to me that this problem could be addressed experimentally by conducting the transformation and subsequent propagation in the absence of selection. If a chimaeric callus is produced as a result of transformation of a protoplast in G_2, the sensitive cells would die if cultivated in the presence of kanamycin. However, both genotypes would survive and could be identified, if maintained without selection.

Potrykus: We have also done many experiments without selection. We split populations after DNA treatment, subjected half to selective conditions, the other half to non-selective conditions, grew those for six weeks and then transferred random clones of the non-selected population to selective conditions and compared the transformation frequency of both—these were identical.

Chaleff: To answer Sir Riley's question, you would have to ask if different frequencies of chimaeras occur among plants regenerated from transformed callus produced in the absence of selection than appear among plants regenerated from callus produced in the presence of selection.

Potrykus: You are correct; we have not searched for chimaeras in non-selected populations. We were excited about this instability in the selected material because we thought that this represented the first chimaera. That was the reason why we used one hundred flowers from this plant to produce offspring covering the entire plant body. However, this plant turned out not to be a chimaera.

Uchimiya: We did a similar experiment and we found no evidence of chimaeras in the transformants. We have done simple Southern blot analysis of the protoplasts to visualize the plasmids. We found that even after one week we still see extrachromosomal plasmids in the protoplasts. However, we used circular plasmids, Dr Potrykus used linearized plasmids. We couldn't see any of the supercoiled form after one week in culture, therefore some kind of nick has been introduced into the extrachromosomal plasmid by this stage.

Galun: So it may be integrated after three days, i.e. it could become integrated during the S phase.

Potrykus: How do you exclude the possibility that these plasmids are adhering to the protoplasts? We can hypothesize about the phase at which integration is occurring but as there are no data at present this is pure speculation.

Cocking: Could I just ask for some background data? Are these leaf protoplasts that you are using?

Potrykus: Yes, this high frequency of up to 10% in tobacco is with leaf protoplasts.

Cocking: From the analyses that people have done over the years it seems that in many tobacco species there is a high synchrony of the cell cycle in the protoplast system; I think most of them are in G_1. That is relevant to the general background of the discussion, that there is every evidence that your protoplast will be in G_1.

Riley: That explains why there are no chimeras.

Galun: Now that it is possible to do this transformation in cereals (monocots), what do you have in mind to transform?

Potrykus: We hope that it is now possible but we have no proof yet. I am open to collaboration and I know there are a number of interesting genes around and a number of labs who would love to put their genes into an homologous system. We don't have our own genes for this purpose.

Hall: Since many people, for example Drs H. Lorz and F. Salamini in Germany have been trying microinjection of microspore-derived tissues for so long without success, perhaps you would like to enlighten us a little as to the secret of your success with this technique?

Potrykus: I do not know what the others may have done wrong.

Hall: If you are tantalizing us with this glorious revelation that the major cereal crops can now be transformed, I feel that you should provide a little evidence.

Potrykus: I have not been claiming that much. I have only been saying that we have clear evidence that injecting a marker gene into microspore-derived pro-embryos of *Brassica napus* produces transgenic chimaeras which can be resolved to solid transgenic plants by secondary embryogenesis (Neuhaus et al 1987). Then I said that by applying exactly the same procedure i.e. microinjecting into as many cells as possible of microspore-derived pro-embryos of wheat, rice, barley and maize, then culturing this material in such a way that it germinates directly, you can produce plants which may or may not contain the gene. These plants are expected (by analogy to *B. napus*) to be transgenic chimaeras, if they are transgenic at all. We have taken leaf tips of a number of such putative transgenic chimaeras for dot blot analysis but this method can produce artifacts or misinterpretations.

Hall: Are these maize leaf tips? You gave specific numbers, such as 4%.

Potrykus: The value of 4% was from maize and the material was from the tips of the first two leaves.

Hall: That's a fairly strong statement—4% of 200 plants is pretty good.

Potrykus: So far we only have dot blot analysis and you can believe or not believe that: we would, of course, like to believe but we are not 100% convinced so we are growing these plants now to produce seeds and producing more transgenic plants by doing more injections.

I would like to stress that we do not claim to have unequivocal evidence for having produced transgenic cereal plants—we have indicative evidence but we prefer to wait until we have seen the analysis of the sexual offspring before we state in public that we have produced transgenic cereals.   Let me give you some more details. G. Neuhaus and G. Spangenberg do exactly the same as with *Brassica*. They take microspore-derived pro-embryos of the size and the cell number optimal for microinjection of DNA, i.e. between 12–16 cells. The procedure is to use a holding pipette and to inject into as many cells as possible, preferably into the nuclei by rotating this pro-embryo at the mouth of the holding pipette. These injected pro-embryos are cultured in little volumes of culture medium. The ratio of culture medium volume to tissue or cell volume is an important parameter, a ratio of about 500:1 to 1000:1 seems to be ideal. We have another talented co-worker, S. Datta, who can obtain further development of these microspore-derived embryos via direct germination.   In *Brassica*, it is possible to produce secondary embryos from epidermal cells of the hypocotl. This has been done to investigate whether microspore-derived injected embryos are chimaeric or not. Individual secondary embryos have been taken and analysed for the presence or absence of the foreign gene by dot blot analysis. This shows that the primary embryo is a chimaera. Subsequent studies with individual secondary embryos have shown that these are solid transformants. The detailed molecular proof for the integration of this foreign gene is in Neuhaus et al (1987). The following is private and preliminary information. We have also obtained cereal pro-embryos derived from microspores; it does not matter which cereal it is, you cannot see a difference in the structure and shape of microspore-derived embryos from cereals and *B. napus*. The technique is exactly the same. Cereal plant cells have denser cytoplasm than *B. napus* cells, so injection into the nucleus is more difficult. However, the transformation frequency by injection into the nucleus is about 80%, the transformation frequency by injection into the cytoplasm is about 40%, so it doesn't make too big a difference.   The rest of the procedure is the same as for *B.napus*. These early pro-embryos are grown in small volumes of culture medium covered by an oil droplet. We feel that the microculture is an important part because not only the injection but also the recovery of the injected pro-embryos is important and this requires microculture. More details are available in the review article by Schweiger et al (1987). You induce further development by the appropriate culture medium and it is advantageous to have someone with 'green fingers', like my co-worker, S. Datta! The further developed embryos are transferred singly to small wells in culture medium. We try to get direct germination and to avoid culture formation. The most developed barley plantlets are now at the 10 cm stage, many more are following. This work was only begun on July 3rd 1987 and in between we have also moved into a new laboratory, which has meant further delays while some of the facilities were finished and equipped. The wheat plants are also about 10 cm, we lost the first set because of an unfortunate mistake. The furthest developed rice plants are about 20 cm high and the maize plants are approximately 30 cm tall. As I mentioned, from dot blot analysis we had the impression that in maize about 4% might contain the foreign gene. For barley, wheat and rice we have not yet performed the dot blots.   I want to repeat what I have stressed several times: these are running experiments, to us promising experiments, but we have no proof that these are transgenic plants. We will examine the offspring and in the meantime produce as many further specimens as we can.

Hall: I am delighted that you showed us pictures and I think it really encourages people that such important crops will be transformed. Can you raise these cultures from any lines? For example are these elite lines you are using? Do you use linear DNA constructions and what sort of genes are you putting in?

Potrykus: We are at the early stages of testing for how many lines we can do this. We have the impression that as long as you have an anther culture system, this technique is applicable. I mentioned that the rice and wheat anther cultures are our own; the work on barley is a collaboration with the Carlsberg laboratory in Copenhagen, and the maize experiments are in collaboration with the Institute of Genetics, Academia Sinica, Beijing.

Hall: There has to be some care, doesn't there, with anther cultures that you get gametophytic tissue not somatic tissue?

Potrykus: Well the ideal situation would be isolated microspore culture, which is not yet possible with cereals except for barley. We had an isolated microspore culture for barley but this will not be easy in wheat, maize or rice. So in these cases we took anthers after some time in pre-culture, opened the anthers and took out the young developing pro-embryos. It is easy to discriminate between microspore-derived pro-embryos and anther wall-derived tissue.

Riley: But the dot blots you showed us of *B.napus* were of haploid plants?

Potrykus: These are haploid plants. But *B.napus* has a high spontaneous doubling rate.

Uchimiya: Are these dot blots probed with DNA or RNA?

Potrykus: DNA.

Komamine: Did you compare microinjecting only one cell in a pro-embryo with injecting several cells?

Potrykus: I do not recall whether this has been done deliberately. I think the experiments were always done with the aim of injecting as many cells as possible.

Komamine: You tested transformation in the whole secondary embryo, at which stage I think it's possible that histogenesis has already occurred. Is it possible that in some tissues the cells are transformed but not in other tissues?

Potrykus: Yes, I think there is proof that this method produces transgenic chimaeras. The dot blot of secondary embryos produced from a single primary embryo showed positive and negative examples. There is no question that the immediate product of microinjection into pro-embryos is a chimaera.

Cocking: You talked about work on isolated protoplasts and direct plasmid interaction with protoplasts and you listed a number of positive features of this approach. You also included some negative features, and amongst those was the range of genotypes, within a given species, that were responsive to whole plant regeneration from protoplasts. Could you draw some parallel, as you see it, between this direct plasmid interaction with protoplasts and injection of plasmid DNA into developing microspores. As I see it, on the one hand there is the technical difficulty associated with injection, which perhaps isn't so difficult but is very demanding of the person; you also have the requirement for a more sophisticated microculture, at least at the early stages in the development. There are also the positive features that in certain of the cereals it is established that you can regenerate plants from anther cultures.

I would like to ask this question: when dealing with a range of genotypes within a species, do you see a pronounced advantage of what I would call your 'transformation by injection', as compared with the use of isolated protoplasts? Particularly since people are currently working on protoplast systems for lots of things other than transformation and therefore these systems are being developed for a range of genotypes (varieties) within the species. There are now at least a dozen varieties of rice, mainly japonicas and perhaps one or two indicas, which can be regenerated from protoplast to plant. So how do you see the comparative merits of these approaches? I know it may be a bit early for the one you described.

Potrykus: It's certainly too early because we have too little experience with the microinjections. I do not intend to give the impression that microinjection into microspore-derived pro-embryos makes work with cereal protoplasts unimportant.

Cocking: No, I didn't say that—I just wanted to get your view on the comparison as you saw it.

Potrykus: If somebody can successfully direct plant regeneration from protoplasts in any cereal, they will not need microinjection into pro-embryos to transfer genetic material. Only time can tell whether direct gene transfer or microinjection will be the method of choice for cereals.

Riley: So far as rice is concerned Ted, it is easier to culture anthers of japonica rices than it is of indica rices. Is it also easier to handle protoplasts of japonica rices?

Cocking: Yes. So you have the exact parallel there and this is the point I was trying to bring out, whether roughly the same sorts of problems are inherent in both approaches. As I see it—both systems are very exciting—both systems depend on getting the DNA into the cells. One method (microinjection) is more technically exacting than the other, but both seem to come down to this common denominator, which is going to be for the breeder, whether you can work on a particular variety of interest. If you can't get that variety to respond, the breeders often lose interest.

Potrykus: It's surprising how long it has taken to produce transgenic rice after these many reports on plant regeneration from rice protoplasts. Even in the rice protoplast regeneration system, I think it might be worth testing microinjection for transformation. There seems to be some problem with transformation in rice. The other point is that what I have shown you is the work of less than three and a half months in a lab which we are still establishing, during which we also ran an EMBO course. What we are looking at is the beginning of the work, not the endpoint, and, of course, we are following some ideas in addition to what I have shown you.

Cocking: If I could add from my experience of the field; this work has built splendidly on the foundation of work of Professor H.U. Koop and his colleagues on the setting up of the microculture system (Koop et al 1983). Until the further stimulus that your work is going to give it, this has been largely ignored by other workers.

Potrykus: That is why I stressed several times that people might be interested to look at the review article by Schweiger et al.

Cocking: I have read it, it's very interesting.

Galun: About variety problems, we shouldn't forget that usually we are not going to inject separately into each variety of a given species. I think that in the future this methodology will be used to produce major changes in cultivars which are difficult to handle in other ways, for example by normal breeding. Therefore, we always have to consider that obtaining a transgenic plant is by no means the end of the breeding process, this is actually only the beginning. It should be integrated, so that under any conditions we will have to crossbreed the transgenic plant with plants of other varieties and complete the breeding programme using an appropriate method of selection. I don't think this is a problem. Once you have an important novel type of plant which expresses a foreign gene in a wheat or rice cultivar, then you can easily crossbreed it with other cultivars of the same group. In rice there may be a difficulty between two types of rice, but within the same type of rice it shouldn't be a problem. This is different from anther culture where in many cases you want to 'purify' the heterozygous type and produce homozygous types, which is a direct breeding problem. There you have to have this methodology available in a variety which you are breeding in order to clean it. The transfer of foreign genes introduces something from the outside, possibly even from another kind of organism or a synthetic gene.

Chaleff: Dr Potrykus, I suspect from my own limited experience with these cultures that the experimental advantage of using microspore-derived embryos for microinjection is that these embryos exist as independent masses that can be readily manipulated. Would you comment on what you perceive to be the advantages of using microspore-derived embryos over embryos present in embryogenic callus lines derived from other sources, such as scutellar tissue?

Potrykus: If we can identify these somatic embryos at an early stage, I would certainly consider it worthwhile to try.

Chaleff: It has been mentioned that the haploidy of the microspore-derived embryo system is a disadvantage. In my work with rice anther culture, I found the frequency of diploidization to be genotype dependent and the regeneration of diploid plants to be far from a trivial matter with particular cultivars.

Potrykus: If somatic embryos are small enough, certainly I would use them for microinjection.

Harms: I would like to reiterate what Professor Galun has said. Plant breeders as well as farmers, if we talk about applications of this type of work, are quite conservative and they want to be absolutely sure that the plants they are going to use in the field are true to type and perform in the field as expected. We are all well aware of the problems, or opportunities as you may call it, of somaclonal variation. As far as our breeders at CIBA-GEIGY are concerned, they are quite critical about the plants we produce. Everything has to pass their stringent evaluation in the field. That imposes on us the requirement to produce a variety of genetically engineered plant materials from which the breeders will ultimately make a selection.

To return to what Professor Cocking said with respect to how many varieties we might need to manipulate, there is always the opportunity to do conventional breeding. Which approach is chosen will depend on their relative economies. Conventional breeding in corn, for instance, takes 5–7 years to convert one genotype into another. Hence, working with A188 because it is a responsive tissue culture system and then trying to breed the novel traits into elite corn inbreds is going to cost you seven years of time and money. You have to balance that against the effort it takes to develop the tissue culture and introduce the foreign traits into each and every elite breeding line that you have. This is a continuous process, if we talk about modifying an elite corn line now, by the time the product reaches the market place, this elite line is going to be old, because the breeder will have made progress developing his germplasm during the interim period.

Scowcroft: You raised one of the issues which a number of us are now facing, namely the speed at which the technology that Ingo (Potrykus) discussed is progressing. The whole area of plant breeding needs complementary technology, so that you don't have to transform every variety independently, rather a single successful transformant within the targeted crop species should be sufficient. *B. napus* is a classic example, where double haploid technology is so good that you can backcross a gene into any variety and within about eighteen months have the equivalent of six generations of backcrossing. It's also essential from the plant breeding and application standpoint to develop not only these transformation technologies that Ingo is working with, but also complementary technologies to overcome the problems that Christian Harms addressed.

Hall: I think Professor Cocking raised a very important point when he asked about the comparison of the two approaches: the protoplast approach where you go through a regeneration of a single cell or groups of cells into which you can introduce a selectable marker; and the groups of cells, for example, that Ingo (Potrykus) showed—I wasn't sure of the number in the cereal system but it looked like he had quite a large number of cells that are not selected. Roughly, how many cells would you say were microinjected?

Potrykus: The pro-embryo I showed from *Brassica* was an ideal size.

Hall: Yes, but are the cereal cells not too small?

Potrykus: It may be possible to use larger pro-embryos, we don't know yet.

Hall: Therefore one is not applying a selectable system and you have to hope that one of the cells that you get in there is indeed transformed. We know that the transformation frequencies are typically not as high as we would want. But may be you have some better data from *B. napus* concerning the frequency of transformation of those cells by microinjection.

Potrykus: The frequency of transformation by microinjection is better studied in single cells. The data I gave you are from single cells, up to 80% if you inject into the nucleus and up to 40% if you inject into the cytoplasm. If you work with a 16-cell pro-embryo the frequency of transformation is between these values. That's the reason for the chimaeras I showed. But there will be different integrations in every single cell, where there is integration, that's why it's important to go to the next generation where meiosis will resolve this complex situation and the offspring will consist of solid transformed clones.

Riley: That's right but of the examples that you described, three of them are inbreeding species, corn is an outbreeder. Were you using an inbred line of corn for generating the microspore-derived embryos?

Potrykus: This is a self-fertile variety.

Riley: So it's an inbred line?

Potrykus: On this question of which method will be the most advantageous, let me describe the situation as I see it. We started on July 3rd 1987 and the protoplast work towards transformation started early in 1986. Let's see who first has convincing data of transgenic cereals—rice, wheat, barley and maize—then we can at least judge which method is faster!

Hall: If you used a construction with the *E. coli* β-glucoronidase gene (GUS) that provides a very sensitive test system, then with the pretty big plant you showed, surely you could confirm activity? If you had a sizeable proportion of your cells transformed with GUS, you would certainly see fluorescence. Is that the construction you are using? That approach is a lot faster than dot blots.

Potrykus: We have already put a number of genes into this material.

Riley: And you are inserting plasmids?

Potrykus: Yes, linearized plasmids.

Scowcroft: What percentage of injected microspore-derived embryos in *Brassica* yield at least one transformed secondary embryo?

Potrykus: As far as I recall, it's better than 80%.

Takebe: Could you comment on the prospects of transforming chloroplasts by your direct gene transfer method, chemically or electrically, or by microinjecting DNA into chloroplasts?

Potrykus: I hesitate to comment about possible developments in the future. At the moment nobody has definite proof for stable integration of a foreign gene into chloroplast DNA. There is some indicative evidence that tDNA may have been introduced into the chloroplast but lost later. We also have genetic data which strongly suggest chloroplast transformation, but there is no proof at the molecular level.

Galun: You are speaking about angiosperms, not algae, not *Chlamydomonas*?

Potrykus: Yes, I would see a good prospect for microinjection into cells which have large chloroplasts.

References

Koop H-U, Weber G, Schweiger H-G 1983 Individual culture of selected single cells & protoplasts of higher plants in microdroplets of defined media. Z Pflanzenphysiol 112:21–34

Neuhaus G, Spangenberg G, Mittelsten-Scheid O, Schweiger HG 1987 Transgenic
    rapeseed plants obtained by the microinjection of DNA into microspore-derived
    embryoids. Theor Appl Genet 74:30–36
Schweiger HG, Dirk J, Koop, HU 1987 Individual selection, culture and manipulation
    of higher plant cells. Theor Appl Genet 73:769–783

Germplasm preservation

Lyndsey A. Withers*

Department of Agriculture and Horticulture, University of Nottingham School of Agriculture, Sutton Bonington, Loughborough, LE12 5RD, UK

Abstract. There is a need for culture storage by both conventional and novel techniques in plant breeding and in the *in vitro* production of secondary compounds. Cultures can be stored by slow growth or cryopreservation. Slow growth is most successful for shoot cultures. Medium-term storage in increments of one or two years can be carried out for several species. Callus cultures are less amenable and there is evidence for progressive deterioration of some traits. Cryopreservation is the only sensible option for long-term storage. Suspension and callus cultures of many species can be cryopreserved. Results on recovery and stability are very encouraging. Embryo cultures show great promise but serious problems are encountered in the cryopreservation of shoot tips. Large differences in response are observed between species, and low survival levels are often compounded by disorganized development. To date, most cryopreservation research has involved an empirical approach but if further progress is to be made, a greater understanding needs to be gained of freezing injury and its control at the molecular, biochemical and cellular levels. *In vitro* conservation should be viewed as one of a range of complementary conservation strategies that range from conventional gene banking to the development of gene libraries.

1988 Applications of plant cell and tissue culture. Wiley, Chichester (Ciba Foundation Symposium 137) p 163–177

A major objective of current plant cell and tissue culture work is the more efficient exploitation of specific properties of plant genotypes. Progress in each of several different areas, for example, the generation of variation, the selective transfer of genes, the identification of desirable traits and the conversion of *in vitro* cultures to plants, contributes to this increased efficiency. However, these efforts are wasted if the products of *in vitro* procedures are prone to variation or, worse, accidental loss. Thus, germplasm preservation must be a component of any *in vitro* programme wherein genetic identity and integrity are important.

Crop improvement by conventional selection and breeding relies upon genebanks both for source material and to conserve improved genotypes. Seed genebanks can provide secure medium- and long-term storage for species that produce orthodox seeds. There are, however, certain crops for which

* *Present address*: International Board for Plant Genetic Resources Headquarters, c/o Food and Agriculture Organization of the United Nations, Via delle Terme di Caracalla, 00100 Rome, Italy

seed storage is unsuitable or impossible (Withers & Williams 1985). Several staple root and tuber crops and a number of tropical and temperate fruits are vegetatively propagated. In some cases the material is sterile, in others gene combinations would be lost by conversion to seed. Such material is stored in field genebanks. These are expensive to run and are susceptible to environmental and pathogenic risks. Crops producing recalcitrant ('short-lived') seeds also have to be conserved at present in field genebanks, as their seeds die after relatively brief periods of storage. Important crops in this category include coconut, cocoa, rubber and mango.

In vitro storage has been proposed for the genetic conservation of these problem crops. Thus, both the conventional and the new, biotechnological approaches to crop improvement have a common need for the development of successful culture storage procedures.

A third area of need can be identified in secondary product synthesis where clonal instability is common. Not only is it practically more sensible for the cultures to be stored without conversion to and from plants, but some of the genotypes in question often only exist *in vitro*.

Serious attempts to develop *in vitro* storage procedures for plants began only about 15 years ago and we are far from having procedures applicable to all species and culture systems. This is due partly to a lack of appropriate research and partly to our poor understanding of storage phenomena. However, the list of subjects that are amenable to storage is growing rapidly and for certain ones routine implementation is possible (see Kartha 1985, Wheelans & Withers 1988, Withers 1985, 1987a,b).

Methods: Technical approaches to germplasm storage in vitro

Slow growth

Cultures can be induced into a state of slow growth by various means, including culture at a reduced temperature or in the presence of growth regulators (natural and synthetic) or osmotically active additives (Staritsky et al 1986, Withers 1987a,b). Restricted nutrient supply, supercooling ('undercooling') and reduced oxygen tension have also been used but with rather less success. Reduced lighting may be advantageous but there has been little systematic investigation of this. In difficult cases, it can help to avoid a storage temperature that is too low, to combine two stresses, e.g. temperature and osmotic, or to warm cultures intermittently (Staritsky et al 1986).

Caution would suggest that slow growth should not be viewed as a sensible long-term option, as some risk of loss must accrue at each subculturing operation and during each storage interval. For truly satisfactory long-term storage, a method is needed that eliminates renewal operations by suspending growth.

Cryopreservation

The term 'cryopreservation' is used exclusively to cover the storage of living biological material at ultralow temperatures (normally at or near the temperature of liquid nitrogen: $-196\ °C$). Except for orthodox seeds, dormant buds and some pollen, higher plant structures cannot generally survive the transition to and from the storage temperature without protection. A typical cryopreservation procedure consists of the following stages: pregrowth, cryoprotection, freezing, storage, thawing and recovery. The exact treatments given at each stage will vary with culture system; details can be found in the literature (Kartha 1985, Sakai 1986, Withers 1985, 1987a,b). Some general points are made here.

Pregrowth provides the opportunity for selecting or inducing the most freeze-tolerant growth phase. For cell suspension cultures, this is late lag or early exponential phase. Pregrowth in the presence of additives such as amino acids, mannitol, sorbitol, sucrose, abscisic acid or dimethyl sulphoxide (DMSO) can enhance freeze-tolerance. During pregrowth, cells grow into a more freeze-tolerant state with changes in cell size, degree of vacuolation and cell wall flexibility. Some treatments appear to select stress-tolerant cells from within a population. For shoot tips, pregrowth provides the opportunity to recover from dissection damage.

Mixtures combining one or both of the classical cryoprotectants, DMSO and glycerol, with a sugar, sugar alcohol, amino acid or polyethylene glycol, are usually employed for cells and calli, whereas DMSO alone is the most frequent cryoprotectant for organized cultures (shoot tips and embryos).

The rate of freezing has dramatic implications for survival. Slow cooling brings about dehydration and an optimum situation between under- and over-dehydration is sought. For many cultures, a rate of $-1\ °C/minute$ or less is effective. Moderately rapid freezing is nearly always lethal; but if the rate is increased further, microscopically small ice crystals will form, which need not be damaging as long as thawing is carried out sufficiently rapidly. Rapid freezing is only effective for certain organized cultures.

Liquid nitrogen-cooled refrigerators provide the only really safe mode of storage. If the storage temperature is allowed to rise, damaging ice recrystallization or, worse, actual thawing can occur.

With few exceptions (see below), thawing is carried out rapidly, again to avoid ice recrystallization damage. The freshly thawed specimen is very vulnerable and must be handled with care. Washing can be very damaging and is best avoided, as is the use of liquid medium. Standard medium formulations are normally used for unorganized cultures. Organized cultures may require additional growth regulators to achieve regeneration.

Results and discussion

Slow growth storage: successes and limitations

Slow growth storage has been extremely successful for shoot cultures of a wide range of species including several root and tuber crops (potato, cassava, yam, sweet potato, aroids), fruits (banana, apple, pear, strawberry, kiwifruit) and other horticultural and agricultural species (Withers 1987b). Both tropical and temperate subjects have been stored, and medium-term storage for up to a decade in increments of one or two years seems to be a reasonable expectation. This represents a significant saving in labour and material requirements. As long as shoot cultures in slow growth remain fully organized, it is unlikely that problems will be encountered with respect to genetic stability. However, morphology is often changed in slow growth and some types of shoot culture with a very compact habit might switch to adventitious development without this being readily apparent.

Attempts to use slow growth for unorganized cultures have been relatively unsuccessful (Hiraoka & Kodama 1984). Some species are able to resume growth fairly readily after storage for up to ten months at a reduced temperature, but others survive for only three months or less. The retention of root forming ability has been demonstrated in slowly grown callus of *Bupleurum falcatum* stored for more than one year, during which time controls lost morphogenic potential. However, the ability to resume secondary product synthesis is seriously impaired in some species. This is probably a direct consequence of depression of the primary metabolism and may be reversible if vigorous growth can eventually be resumed. In general, the possibility of using slow growth for unorganized cultures is viewed cautiously, as it involves the unsatisfactory combination of a culture system prone to somaclonal variation and the risk of the selection of variants under stress conditions.

Cell and callus cultures: candidates for routine cryopreservation

The cryopreservation requirements of cell suspension and callus cultures are very well understood and several laboratories now routinely store cultures in liquid nitrogen (Withers 1985, 1987b, Withers & King 1980). High survival rates of 50–80% may be expected with, in the best cases, cell division resuming within two to four days of thawing. Some species are problematic but these are usually difficult to handle in normal culture. Results on stability in cryopreservation are good; mutant cell lines showing antimetabolite resistance and those capable of synthesizing certain secondary products have been shown to be stable (Seitz et al 1983, Withers 1985, 1987b).

Callus cultures are more difficult to cryopreserve, probably because of the greater bulk of the cell aggregates and the presence of larger, senescing cells. However, aggregates can be cut into small pieces before cryoprotection and then treated as cell suspensions. Stability of desirable traits has again been demonstrated (Watanabe et al 1985, Ziebolz & Forche 1985).

Cryopreservation of shoot cultures: problems of survival and recovery

Shoot cultures have received as much attention as cells and calli but have proven far more variable in response and, generally, more difficult to cryo-preserve (Kartha 1985, Sakai 1986, Withers 1985, 1987a,b). The greatest successes have been achieved with dormant buds derived from trees. Other material originating *in vivo*, e.g. shoot tips dissected from established plants or seedlings, is more difficult to handle but is still more responsive than counterparts produced *in vitro*. Strong intergenotypic variations are seen and poor quantitative survival is compounded by poor quality recovery.

Shoot tips are often the specimens of choice for conservation, as they are thought to be intrinsically genetically stable. However, if freezing damage is severe, this advantage may be lost. Direct regeneration is by no means impossible but it is common practice to increase the percentage recovery by supplementing the recovery medium with growth regulators, leading to cal-lusing and adventitious regeneration (Withers 1985, 1987b). For good con-servation, it may be necessary to avoid this and sacrifice quantity for quality.

The cryopreservation of shoot tips is further complicated by the fact that survival can be achieved by both rapid and slow freezing. In the case of strawberry, carnation, potato and oilseed rape, both mechanisms are effec-tive. In the latter two cases at least, it appears that slow freezing is more likely to lead to adventitious regeneration. Despite there being little quantitative difference between the two situations, slow freezing seems to cause a diffe-rent type of damage, probably involving greater disruption of the symplasm by breakage of plasmodesmatal connections between adjacent cells (L.A. Withers & E.E. Benson, unpublished work 1987). These findings suggest that it may be worthwhile to try the technique of vitrification for shoot tips (see Withers 1987b).

A thread of continuity and logic can be seen in efforts over the years to develop cryopreservation procedures for cell cultures but for shoots it is difficult to build upon published findings. This is perhaps not surprising, as shoots are heterogeneous in so many respects, not least in their basic *in vitro* culture requirements. However, it is possible to take a systematic approach to developing a cryopreservation procedure for an untried specimen and 'home in' on a successful combination of treatments. The author and colleagues have done this very successfully, most recently with shoots of pear (W.H. Wanas & L.A. Withers, unpublished work 1986).

Embryo cultures: 'natural' systems for cryopreservation

Although the cryopreservation of relatively few species has been explored, some notable successes have been scored, leading to the suggestion that somatic embryos may be more naturally predisposed to cryopreservation than shoots or unorganized cultures. Somatic embryos of carrot can be recovered at a very high frequency, developing non-adventitiously after 'dry freezing' (Withers 1979). The cryoprotected specimen is drained of superficial moisture, enclosed in a foil envelope, frozen slowly and then thawed relatively slowly. Increasing dehydration by desiccation before freezing increases the benefit of slow thawing. Somatic embryos of oilpalm pregrown in the presence of a high level of sucrose can be cryopreserved without further cryoprotection and will survive rapid or slow freezing (Engelmann et al 1985, F. Engelmann, personal communication 1986). These findings suggest an intrinsically higher level of freeze-tolerance that may involve physiological factors but may also relate to morphology. Embryogenic systems can regenerate from much smaller units (proembryos and early stage embryos) than can the typical shoot.

Some of the encapsulation methods and pre-treatments used to increase the storage life of artificial seeds may be relevant to cryopreservation (Kitto & Janick 1985). Conversely, the incorporation of cryopreservation into artificial seed handling procedures might resolve the logistical problem of how to store embryos between encapsulation and delivery.

Seeds and zygotic embryos of several species with orthodox behaviour can be cryopreserved without difficulty and some success has also been achieved in the cryopreservation of embryos of recalcitrant seeds (Grout 1986, Normah et al 1986, Withers 1987a). The induction of somatic embryogenesis to reduce the unit embryo size, which can be massive in recalcitrant species, could improve freeze-tolerance.

The nature, consequences and control of freezing damage

The empirical approach has dominated plant cryopreservation research. With technical innovations such as dry freezing (Withers 1979) and droplet freezing (Kartha 1985), it has led to notable progress but it has limitations. Consideration should also be given to understanding the pathology of the injured cell. Each stage in the cryopreservation procedure involves potential sources of stress and injury. Promotion of repair is an important objective but it should be noted that lethal damage can, paradoxically, be less consequential than sub-lethal damage, particularly in terms of genetic stability.

In pregrowth, the stress imposed by osmotically active medium additives is beneficial in that 'hardening' against other stresses (e.g. cold, solute concentration) appears to take place. However, selection by genotype may occur in heterogeneous cell populations, as a result of either the preferential growth of one component genotype or the death of another. There is evidence for shifts in ploidy distribution after pregrowth in the presence of amino acids (Volkova et al 1984). Some pre-treatments may influence the cell cycle and lead to synchronization (Withers 1978).

Cryoprotectant toxicity is a well recognised problem that is difficult to resolve totally. The use of very pure stocks (especially of DMSO) can help where contaminants are the cause. Otherwise, overadministration should be avoided and cryoprotectant mixtures used to give a high net protective effect without additive toxic effects. There is evidence for interference by DMSO with microfibrils, microtubules and the cell division process (see Withers 1987b). There is, fortunately, no evidence for mutation as such but DMSO has been implicated in gene activation in animal and microbial systems (see Benson 1987). The beneficial effects of cryoprotectants include stabilization of membranes and macromolecules through various mechanisms, the most significant of which may be free radical scavenging (Benson 1988, Benson & Withers 1987).

Freezing involves damage resulting from dehydration and ice formation. Membranes suffer structural and functional lesions and significant amounts of membrane surface area may be lost, predisposing the cell to expansion-induced injury (deplasmolysis) on thawing. The storage phase is in many respects extremely stable, as long as the storage temperature is maintained but there is a low risk of cumulative radiation damage (see Benson 1988). Fortunately, some cryoprotectants, including DMSO, are also radioprotectant (protect against radiation).

Thawing is a source of damage in two respects: firstly, it can cause deplasmolysis injury, and secondly it reveals the consequences of damaging events occurring during freezing and storage. Control of the early post-thaw environment provides the opportunity to contain injury and direct cellular processes towards recovery. Evidence from animal systems suggests that manipulation of specific ionic concentrations and ratios may be important, as may the maintenance of ATP levels (see Withers 1987). Culture in the dark or under minimal lighting is likely to help through reducing photooxidation. There is evidence that the plasmalemma of the freshly thawed plant cell is physically weak and the cell is leaky, with depleted pools of key metabolites and a respiratory lesion that takes several days to heal (Cella et al 1982). The latter is a sign of mitochondrial damage, which may lead to oxidative stress and further serious metabolic injury (Benson 1988).

Throughout the early post-thaw period, the cell is likely to be prone to free radical-mediated injury. Evidence gained by measurement of volatile hydrocarbons evolved by recovering cells indicates that membrane breakdown continues for several days, as does free radical scavenging by DMSO. This is a further reason for avoiding post-thaw washing, as the cryoprotectant appears to have an important role in the recovering cell. Evidence from electron microscopy indicates that membrane breakdown, recycling and repair may continue for some time after thawing (Withers 1985). Free radicals may have further important consequences for the stability of cryopreserved cells, as there is evidence for DNA damage in animal and microbial systems (see Benson 1988).

Little is known about the fate of cells interrupted in mitosis, cytokinesis or even in interphase stages of the cell cycle. 'Stage sensitivity' to mutagenesis has been suggested for plant cells, with loss of control of the cell cycle being an important possible source of variation (Gould 1984). The only information available concerning the plant cell cycle and cryopreservation is that cells which have recently passed through mitosis are least prone to freezing injury (Withers 1978).

The consequences of impairment of cell division are less significant for cell suspensions and calli than for organized structures where cell–cell communication and the timing and orientation of cell division are crucial to normal development. Therefore it is not surprising that the resumption of normal shoot development is exceptional. It may even be an unrealistic objective, suggesting that more effort be placed upon embryogenic systems (see earlier section).

There may be a risk of selection for tolerance to freezing or the preferential recovery of certain population components. Populations of mixed ploidy show the same composition before and after cryopreservation (Volkova et al 1984). However, repeated freezing and thawing of callus of *Lavandula vera* does lead to progressively increased survival levels, although there is no corresponding progressive change in secondary metabolite synthesis or morphogenic behaviour (Watanabe et al 1985). It is possible that the increased freeze-tolerance may relate to a morphological change rather than a genetic factor. Despite the general evidence that cryopreservation affords genetic stability, the answers obtained in investigations will only be as discriminating as the techniques used. Current studies using biochemical and nucleic acid techniques may reveal changes that have so far eluded discovery.

Complementary genetic conservation strategies

The storage of cultures will undoubtedly play an important part in the conservation of crop genetic resources in the future. However, it is emphasized that *in vitro* conservation is just one approach among several, to be used only when its adoption is logical. For much material, conventional conservation methods will remain adequate and, at the other end of the spectrum, certain cases will benefit from the development of conservation methods involving isolated nucleic acids, i.e. gene libraries (Peacock 1987).

Storage of cultures is one of a range of *in vitro* procedures that can increase the efficiency of plant genetic conservation (Withers 1987b). Efforts are also being devoted to the development of field collecting techniques and quarantine and disease indexing methods based on culture procedures. The exchange of germplasm in the form of cultures is already a widespread practice (see Wheelans & Withers 1988). Molecular, biochemical and *in vitro* culture techniques are providing new ways of efficiently characterizing and evaluating germplasm, thereby increasing its potential value in crop improvement. *In vitro* techniques will, of course, have their greatest impact in germplasm utilization, be it in the seemingly humble but immensely practical area of mass propagation or the more dramatic area of genetic engineering. However, in our enthusiasm for these new approaches, we should never forget that they will depend upon the provision of desirable genes by sound genetic conservation.

References

Benson EE 1988 Free radical and biochemical damage in stored plant germplasm. International Board for Plant Genetic Resources, Rome, in press

Benson EE, Withers LA 1987 Gas chromatographic analysis of volatile hydrocarbon production by cryopreserved plant tissue cultures: a non-destructive method for assessing stability. Cryo Lett 8:35–46

Cella R, Colombo R, Galli MG, Nielsen E, Rollo F, Sala F 1982 Freeze-preservation of rice cells: a physiological study of freeze-thawed cells. Physiol Plant 55:279–284

Engelmann F, Duval Y, Dereuddre J 1985 Survie et prolifération d'embryons somatiques de Palmier à huile (*Elaeis guineensis* Jacq.) après congélation dans l'azote liquide. Compte Rendu Acad Sci (Paris) 301 Sér. III:111–116

Gould AR 1984 Control of the cell cycle in cultured plant cells. CRC Crit Rev Plant Sci 1:315–344

Grout BWW 1986 Embryo culture and cryopreservation for the conservation of genetic resources of species with recalcitrant seed. In: Withers LA, Alderson PG (eds) Plant tissue culture and its agricultural applications. Butterworth, London, p 303–309

Hiraoka N, Kodama T 1984 Effect of non-frozen cold storage on the growth, organogenesis and secondary metabolism of callus cultures. Plant Cell Tissue Organ Cult 3:349–357

Kartha KK (ed) 1985 Cryopreservation of plant cells and organs. CRC Press, Boca Raton, Florida

Kitto SL, Janick J 1985 Hardening treatments increase survival of synthetically-coated asexual embryos of carrot. J Am Soc Hortic Sci 110:283–286

Normah MN, Chin HF, Hor YL 1986 Desiccation and cryopreservation of embryonic axes of *Hevea brasiliensis* Muell. Arg. Pertanika 9:299–303

Peacock WJ 1988 Molecular biology and genetic resources. In: Brown ADH et al (eds) The use of crop genetic resources collections. Cambridge University Press, Cambridge, in press

Sakai A 1986 Cryopreservation of germplasm of woody plants. In: Bajaj YPS (ed) Biotechnology in agriculture and forestry. Vol 1: Trees 1. Springer, Berlin, p 113–129

Seitz U, Alfermann AW, Reinhard E 1983 Stability of biotransformation capacity in *Digitalis lanata* cell cultures after cryogenic storage. Plant Cell Rep 2:273–276

Staritsky G, Dekkers AJ, Louwaars NP, Zandvoort EA 1986 *In vitro* conservation of aroid germplasm at reduced temperatures and under osmotic stress. In: Withers LA, Alderson PG (eds) Plant tissue culture and its agricultural applications. Butterworth, London, p 277–283

Volkova LA, Popov AS, Samygin GA 1984 Effect of amino acids on a *Dioscorea deltoidea* cell suspension culture and its regeneration after storage at −196 °C. Fiziol Rast (Mosc) 31:632–638

Watanabe K, Yamada Y, Ueno S, Mitsuda H 1985 Change of freezing resistance and retention of metabolic and differentiation potentials in cultured green *Lavandula vera* cells which survived repeated freeze-thaw procedures. Agric Biol Chem 49:1727–1731

Wheelans SK, Withers LA 1988 The IBPGR *In Vitro* Conservation Databases. In: Proc NATO Adv Stud Inst Plant Cell Biotechnol, Albufeira, Portugal. Springer, Berlin, in press

Withers LA 1978 The freeze preservation of synchronously dividing cultured cells of *Acer pseudoplatanus* L. Cryobiology 15:87–92

Withers LA 1979 Freeze preservation of somatic embryos and clonal plantlets of carrot (*Daucus carota* L.). Plant Physiol 63:460–467

Withers LA 1985 Cryopreservation of cultured cells and meristems. In: Vasil IK (ed) Cell culture and somatic cell genetics of plants. Vol. 2: Cell growth, nutrition, cytodifferentiation and cryopreservation. Academic Press, Orlando, Florida, p 253–316

Withers LA 1987a Low temperature preservation of plant cell, tissue and organ cultures for genetic conservation and improved agricultural practice. In: Grout BWW, Morris JG (eds) The effects of low temperatures on biological systems. Edward Arnold, London, p 389–409

Withers LA 1987b Long-term preservation of plant cells, tissues and organs. In: Miflin BJ (ed) Oxford surveys of plant molecular and cell biology. Oxford University Press, Oxford, Vol 4, p 221–272

Withers LA, King PJ 1980 A simple freezing unit and cryopreservation method for plant cell suspensions. Cryo Lett 1:213–220

Withers LA, Williams JT 1985 Research on long-term storage and exchange of *in vitro* plant germplasm. In: Biotechnology in international agricultural research. International Rice Research Institute, Manila, p 11–24

Ziebolz B, Forche E 1985 Cryopreservation of plant cells to retain special attributes. In: Schäfer-Menuhr A (ed) *In vitro* culture – propagation and long term storage. Nijhoff/Junk for CEC, Dordrecht, p 181–183

DISCUSSION

van Montagu: You mentioned that on repeated freezing, the cultures can become more resistant to freezing. You suggested that there might even be genotypic change. Has anyone ever regenerated this material to see if the resistance became an integral genetic trait of the cells or whether it was really only, say, an adaptation of the lipids.

Withers: To my knowledge no one has taken material that far after freezing.

We have only recently developed adequate freezing methods to design the experiments that enable us to do the necessary analysis. I know of at least two projects underway at the moment that are using biochemical and molecular techniques to analyse material recovered from cryopreservation but we don't have the answers as yet, I think that will take two or three years. However, in microbial and animal cell work there is no evidence for genetic instability as a result of freezing injury (Ashwood-Smith 1985), although there is a slight suggestion of gene activation by DMSO. In higher plants, there is definite evidence for intergenotypic differences in freeze-tolerance, which could lead to selection from within mixed populations (Engelmann 1976, Withers 1987 and references therein), but as for actual genetic modification by freezing in plants, we don't know yet.

Yamada: One of the post-doctoral fellows in our laboratory, Dr Kazumi Watanabe, has shown that biotin content in cultured *Lavandula vera* cells is not changed by repeated freezing. We analysed the cell membrane and found that the rate of exchange between saturated and unsaturated fatty acids had been increased by repeated cryopreservation. We thought that this kind of repeated selection by freezing might have changed the constituents of the membrane but the amount of fatty acids, phophatidyl choline and phosphatidyl ethanolamine in the membrane did not differ from the amounts in an untreated cell. We think that the speeds of conversion from saturated to unsaturated fatty acids in the original unselected cells and in the cells which are resistant to freezing are different (Watanabe et al 1985). So I think that there is no need to worry about inducing genetic changes by repeated freezing and thawing.

Harms: You have quite appropriately stressed the problems that are posed by selection when you go into some sort of storage and come back from storage. It may be appropriate in some cases to use selection as an opportunity rather than an obstacle to recover kinds of cells that one wishes to enrich for. I am thinking of examples that could help Professor Komamine in his work, where after the storage of a heterogeneous suspension the recovery favours embryogenic cells over cells that are more vacuolated. So this may be one approach that you could use to enrich for certain types of cells. This approach has been used successfully in order to get more embryogenic types of cells.

Durzan: This is what happens to the gymnosperm cell suspension cultures, one is essentially selecting for embryogenic lines.

I should also say that cryopreservation is a matter of perspective. In boreal forests, where the temperature goes down to -55 degrees Fahrenheit for a good period of the year, much of the germplasm we are interested in is naturally cryopreserved and every year the trees still look like trees! So I am not too concerned about the damage that may be caused by cryopreserving appropriate types of material. The danger is in the abuse the cells get in the pre-treatments.

Withers: There are several mechanisms in nature for avoiding low temperature injury, including freezing avoidance (Sakai 1982). However, I take the point about deep winter freezing, although that involves dormant material in which the cells are not actively dividing. This contrasts with the common situation in culture when cells risk being arrested part way through mitosis. You are describing a very special situation but it may, nonetheless, provide a model for improving freeze-tolerance. Thus, inducing dormancy before freezing may be very useful.

Durzan: This was what one was leading to in terms of pre-conditioning the cells.

Rhodes: Dr Withers, do you have any experience of applying these cryopreservation techniques to root tips?

Withers: It has been done inadvertently when freezing very large carrot plantlets, the products of somatic embryos (Withers 1979). The root tips survive at a frequency similar to the shoot tip.

Rhodes: Presumably, with most work you would be able to increase the frequency of survival of such organized structures after frozen storage.

Withers: We would love to try it!

Galun: You mentioned that treatment with abscisic acid (ABA) is sometimes effective; this is one way of putting cells into dormancy.

Withers: Yes, ABA has been used both in slow growth, e.g. in potato (Westcott 1981) and in pregrowth treatments before cryopreservation (see Withers 1987).

Zenk: We have some experimental evidence from when we froze cell cultures that were high producers of secondary compounds, for example, anthraquinones in *Morinda* or *Galium* (Rubiaceae). We found that these could not survive the freezing process. Strains derived from these same species which did not produce these secondary compounds, were able to be frozen and revived. Has anybody ever looked at what happens to the vacuole during the freezing process? Is there a possibility that the vacuole sap comes into contact with cytoplasm during the formation of the ice crystal? Is there an opportunity for biochemical interaction between secondary phenolic compounds and the cytoplasm?

Withers: Yes, membranes do become extremely leaky even in the potentially surviving cells (Cella et al 1982, Palta et al 1982), so the risk of internal toxins coming into contact with regions of the cell that they can damage is high. Kartha and colleagues (Chen et al 1984, Kartha et al 1982) have attempted the cryopreservation of relatively high and low secondary metabolite producing cells of *Catharanthus roseus* and they have found that both are freezable, but they require different cryopreservation procedures, so don't give up hope yet.

Zenk: But the situation is different—the cultures that I was talking about produce gram quantities of secondary products per litre, the *Catharanthus* cells that Kartha used were producing much less than 100 mg/l. So there is a large quantitative difference.

Withers: But if you know how to increase the production of secondary compounds by those cells, you presumably know how to decrease it by modifying the culture conditions. Perhaps those culture conditions would be suitable for pregrowth before freezing.

Zenk: We tried all sorts of conditions, but in vain.

Fowler: We have had the same experience with freezing as a means of storage as Professor Zenk. The vacuolar contents of the cells froze and then expanded as the cells were thawed. This resulted in massive disruption of the tonoplast and mixing of the cell contents. We have solved this problem by storing callus tissues under glycerol. This involves covering the callus with a layer, about 1–2 cm deep, of glycerol and preferably keeping the cells in a cool environment (+4 degrees Centigrade). Samples of the callus can be removed at intervals and brought back fairly rapidly to full growth. We presume that the system works by reducing the amount of oxygen available to the cells and thereby limiting growth. I am not sure, however, that all cell cultures would remain viable under glycerol; the general approach may be applicable but with the use of other reagents as the submersing fluid.

Yamada: I think this is a very important point. We usually use a combination of DMSO and glycerol as the cryoprotectant. But although DMSO is toxic to plant cells, it increases transpiration of water from the internal cell sap to the outside. We don't know how DMSO works; I believe that DMSO must alter the physical nature of the cell membrane. Mitsui Petrochemical Industries have applied for a patent for a fusion method that uses DMSO as the accelerator of fusion. I think that this works because the structure of the cell membrane has been modified by the DMSO.

Withers: Concerning DMSO, we have found and some scientists in Germany (Matthes & Hackenselner 1981) have found, that DMSO itself has a relatively low toxicity as long as very pure stocks are used, purified, if necessary, just before use. It is possible to purchase extremely pure DMSO, beyond analytical grade ('Spectrograde', for UV spectroscopy). This should be purchased in very small amounts, e.g. 100 ml bottles or even sealed ampoules of smaller volume, and stocks should be kept at room temperature and turned over very quickly. Toxicity may be due to breakdown products that can accumulate within a short period of time.

Potrykus: I can see that the vacuole is a major problem. Has anyone tried to work with meristematic cells, either by specific treatment or even by evacuolation?

Withers: You can reduce the vacuolar volume by rapid subculturing and very heavy plasmolysis (Pritchard et al 1985, Withers 1987 and references therein). Both of those techniques can be effective but the work has to be done on a species by species basis as there are inter-genotypic differences in response to such pregrowth treatments.

Harms: You have been concerned primarily with cold storage, how about dessication storage similar to normal seed storage? There are cases where somatic embryos have been desiccated accidently and then it has been found that they rehydrate pretty well after storage at room temperature, not a low temperature. What are the prospects in that area?

Withers: I know of two examples of the use of desiccation: Nitzsche (1978) and Jones (1974) have tried this with callus and somatic embyros of carrot. The former example also involved freezing to -80°C. However, carrot is a wonderful model sometimes and a terrible model at other times; it will do almost anything you ask it to do, therefore it does not necessarily help develop more widely applicable methods. It would be very interesting to look at desiccation specifically of somatic embryos. That was my point in mentioning artifical seed production because the combination of that technology with desiccation and freezing could be extremely useful (Kitto & Janick 1985, Redenbaugh et al 1986). It may be the solution to the conservation problem of recalcitrant seeds (Normah et al 1986, Withers & Williams 1985).

References

Ashwood-Smith MJ 1985 Genetic gamage is not produced by normal cryopreservation involving either glycerol or dimethyl sulphoxide: a cautionary note, however on possible effects of dimethyl sulphoxide. Cryobiology 22:427–433

Cella R, Colombo R, Galli MG, Nielsen E, Rollo F, Sala F 1982 Freeze preservation of rice cells: a physiological study of freeze-thawed cells. Physiol Plant 55:279–284

Chen THH, Kartha KK, Leung NL, Kurz WGW, Chatson KB, Constabel F 1984 Cryopreservation of alkaloid producing cells of periwinkle (*Catharanthus roseus*). Plant Physiol 75:726–731

Englemann F 1986 Cryoconservation des embryons somatiques de Palmier à huile (*Elaeis guineensis* Jacq.). Mise au point des conditions de survie et de reprise. Doctoral thesis, University of Paris VI

Jones LH 1974 Long term survival of embryoids of carrot. Plant Sci Lett 2:221–224

Kartha KK, Leung NL, Gaudet LaPrairie P, Constabel F 1982 Cryopreservation of periwinkle, *Catharanthus roseus* cells cultured *in vitro*. Plant Cell Res 1:135–138

Kitto SL, Janick J 1985 Hardening treatments increase survival of synthetically-coated asexual embryos of carrot. Soc Hortic Sci 110: 283–286

Matthes G, Hackenselner KD 1981 Correlations between purity of dimethyl sulphoxide and survival after freezing and thawing. Cryo Lett 2:389–392

Nitzsche W 1978 Erhaltung der Lebensfaehigkeit in getrocknetem Kallus. Zeitschrift fur Pflanzenphysiologie 87:469–472

Normah MN, Chin HF, Hor YL 1986 Desiccation and cryopreservation of embryonic axes of *Hevea brasiliensis* Muell. Arg. Pertanika 9:299–303

Palta JP, Jensen KG, Li PH 1982 Cell membrane alterations following a slow freeze thaw cycle: ion leakage, injury and recovery. In: Li PH, Sakai A (eds) Plant cold hardiness and freezing stress vol 2: mechanisms and crop implications. Academic Press, New York, p199–209

Pritchard HW, Grout BWW, Short KC 1986 Osmotic stress as a pregrowth procedure for cryopreservation. 3. Cryobiology of sycamore and soybean cell suspensions. Ann Bot (Lond) 57:379–387

Redenbaugh K, Paasch BD, Nichol JW, Kossler ME, Viss PR, Walker KA 1986 Somatic seeds: encapsulation of plant embryos. Biotechnology 4:797–810

Sakai A 1982 Extraorgan freezing of primordial shoots of winter buds of conifer. In: Li

PH, Sakai A (eds) Plant cold hardiness and freezing stress, vol 2: mechanisms and crop implications. Academic Press, New York p199–209

Watanabe K, Yamada Y, Ueno S, Mitsuda H 1985 Change of freezing resistance and retention of metabolic and differentiation potentials incultured green *Lavandula vera* cells which survived repeated freeze-thaw procedures. Agric Biol Chem 49:1727–1731

Westcott RJ 1981 Tissue culture storage of potato germplasm. 2. Use of growth retardants. Potato Res 24:343–350

Withers LA 1979 Freeze preservation of somatic embryos and clonal plantlets of carrot (Daucus carota L.). Plant Physiol 63:460–467

Withers LA 1987 Long-term preservation of plant cells, tissues and organs. In: Miflin BJ (ed) Oxford surveys of plant molecular and cell biology. Oxford University Press, Oxford, Vol 4, p 221–272

Withers LA, Williams JT 1985 Research on long-term storage and exchange of *in vitro* plant germplasm. In: Biotechnology in international agricultural research. International Rice Research Institute, Manila p11–24

Elicitation and metabolism of phytoalexins in plant cell cultures

W. Barz, S. Daniel, W. Hinderer, U. Jaques, H. Kessmann, J. Köster, C. Otto and K. Tiemann

Lehrstuhl für Biochemie der Pflanzen, Westfälische Wilhelms-Universität, Hindenburgplatz 55, 4400 Münster, Federal Republic of Germany

Abstract. Application of biotic or abiotic elicitors to plant cells induces substantial metabolic alterations directed at establishing plant defence reactions. The elicitor-induced accumulation of antimicrobial phytoalexins deserves special attention for explaining plant-fungal parasite interaction. The great chemical diversity of phytoalexins is reviewed. In chickpea (*Cicer arietinum* L.), the 6aR:11aR-pterocarpan phytoalexins, medicarpin and maackiain, are induced by both endogenous and *Ascochyta rabiei*-derived elicitors. An *A. rabiei* suppressor inhibits in *C. arietinum* the accumulation of pre-infectional isoflavones, their glucoside conjugates and the phytoalexins. Chickpea cell cultures established from cultivars with high resistance (ILC 3279) and high susceptibility (ILC 1929) to the pathogen *A. rabiei* show identical patterns of isoflavone accumulation but differ significantly in phytoalexin production. The high phytoalexin producing culture ILC 3279 has been used to characterize new isoflavone hydroxylases and an isoflavone oxidoreductase which are specifically involved in pterocarpan formation. The elicitor-induced changes in enzyme activities measured in cell culture ILC 3279 can be depicted by a metabolic grid of three sets of closely linked enzymes for the general phenylpropanoid pathway, the isoflavone conjugation reactions and the biosynthesis of pterocarpans. After excretion into the growth medium, the pterocarpans are polymerized by peroxidases.

1988 Applications of plant cell and tissue culture. Wiley, Chichester (Ciba Foundation Symposium 137) p 178–198

Plants have developed sophisticated mechanisms to combat stresses exerted on them by a vast range of environmental agencies and to defend themselves against pathogenic or non-pathogenic microorganisms. In addition to various preformed static barriers of materials such as cutin or suberin and constitutively produced protective compounds ('pre-infectional inhibitors'), plants possess a range of active defence mechanisms which are started as a consequence of physical damage, disorganization of tissue or by perception of a chemical signal provided by an attacking microorganism. Among the various defence reactions, the induced *de novo* synthesis and rapid accumulation of low molecular weight, antimicrobial compounds ('phytoalexins') (Bailey &

Mansfield 1982) have attracted special attention. The differential accumulation of these defence compounds in incompatible and compatible plant–pathogen interactions plays a crucial role in the specificity of host resistance (Wood 1986, Dixon 1986).

Elucidation of the biochemical and molecular biological reactions involved in microbial attack of plants cells, in the spreading of infection and in the development of the induced plant responses will clarify the specific role of both partners in the interaction. Such investigations will also reveal the chemical structure of biological signal compounds, as provided by the attacking microorganisms, and the mechanisms of signal perception and transduction in the plant cells. Furthermore, such knowledge will greatly contribute to modern breeding programmes to produce crop plants less prone to parasitic attack.

Plant cell suspension cultures of host and non-host plants have been widely used as excellent experimental models for the study of induced plant defence mechanisms and the reactions involved in phytoalexin formation (Dixon 1986, de Wit 1986, Kombrink et al 1986, Ryder et al 1986, Barz et al 1987). When compared to intact plant tissues, cell cultures represent systems of reduced complexity. Since in cultures the number of responding cells is much higher, these cell cultures greatly facilitate initial biochemical analyses, measurements of the rapid responses of plant cells, ready detection of *de novo* synthesized compounds, purification of enzymes and isolation of mRNA for cDNA cloning. The advantages and disadvantages of using cell suspension cultures as a model system in initial biochemical studies of phytopathological interactions have been discussed (Dixon 1980, Kessmann & Barz 1987).

Elicitation of phytoalexins and other defence reactions in plant cell cultures

Accumulation of phytoalexins ('post-infectional inhibitors') in cell cultures or infected plant tissues is a result of increased synthesis from remote precursors (Dixon 1986) and major increases in the levels of appropriate biosynthetic enzymes are characteristically observed at the onset of the defence response (Dixon 1986, Ebel 1986). All available evidence indicates that these biosynthetic enzymes are formed by *de novo* synthesis and that they are induced after perception of an extracellular signal. Phytoalexin formation thus follows the principle of differential gene activation of appropriate biosynthetic enzymes (Ward 1986, Ebel 1986).

All molecules leading to an induction of phytoalexins in plant cells are summarized under the term 'elicitors' (Ward 1986). Various types of 'biotic elicitors' (polysaccharides, proteins, glycoproteins, unsaturated fatty acids) stem from the cell walls, culture filtrates or cytoplasm or parasitic, as well as non-parasitic, bacteria and fungi (de Wit 1986, Darvill & Albersheim 1984).

The chemical structures of glucan elicitors from *Phytophthora megasperma* f. sp. *glycinea* cell walls, mainly tested on glyceollin accumulation in soybean cells, have been elucidated (Darvill & Albersheim 1984). Fragments of cell wall pectin ('endogenous elicitor') were also shown to be effective as an elicitor. Such oligo-α-1,4-D-galacturonides are thought to be formed either by plant hydrolases liberated upon wounding of plant tissues or by fungal polygalacturonases (Darvill & Albersheim 1984, de Wit 1986). Most of the elicitors, when applied to cell cultures, do not simulate the specificity normally not even show species specificity (de Wit 1986).

'Abiotic elicitors' include UV light, heavy metal ions, detergents and various xenobiochemicals, as well as freezing or heating of plant cells. These elicitors, when applied to cell cultures, do not stimulate the specificity normally found in host-parasite interactions. However, biotic and abiotic elicitors show an important difference with regard to dose-dependent induction of phyto-alexins. In studies with chickpea (*Cicer arietinum* L.) cell suspension cultures and their pterocarpan phytoalexins medicarpin and maackiain (for structures see Fig. 5), the dose dependence of biotic elicitors resulted in saturation curves of phytoalexin production, whereas heavy metal ions (Cu^{2+}, Hg^{2+}, Mn^{2+}) exhibited pronounced optimum curves, indicating strong toxin effects of such abiotic elicitors at higher concentrations (U. Jaques & W. Barz, unpublished work 1987).

Application of various elicitors to different plant cell cultures leads to substantial changes in cellular metabolism. Fig. 1 presents a scheme of the main metabolic alterations observed so far. They comprise effects on Ca^{2+} ion metabolism (Kauss 1987), differential gene activations leading to the *de novo* syntheses of enzymes involved in the formation of polysaccharides (e.g. callose), hydroxyproline-rich glycoproteins in cell walls by induction of prolyl hydroxylase, lignin and polyphenols similarly deposited in cell walls, chitinase, proteinase inhibitors, pathogenesis-related (PR) proteins and phytoalexins. Furthermore, massive changes in membrane integrity, respiration, protein and phosphate metabolism, ethylene production or peroxidase activity have also been found to occur (Dixon 1986, Ebel 1986, Ward 1986). The involvement of plant cell cyclic AMP in transducing the elicitor signal has been discussed controversially (Kurosaki et al 1987). These results on elicitor action indicate that plant cells employ a considerable number of pathways in defence reactions against microorganisms.

Numerous phytoalexins from plants and plant cell cultures of many different families are known. Though a great variety of chemical structures has been found (Fig. 2), phytoalexins may be useful chemotaxonomic markers because certain carbon skeletons are characteristic for a particular genus or family (e.g. sesquiterpenoids /*Solanaceae*; isoflavonoids /*Leguminosae*; dihydrophenanthrene /*Orchidaceae*). More than 200 phytoalexins have been identifed and due to continued efforts in many laboratories the number of

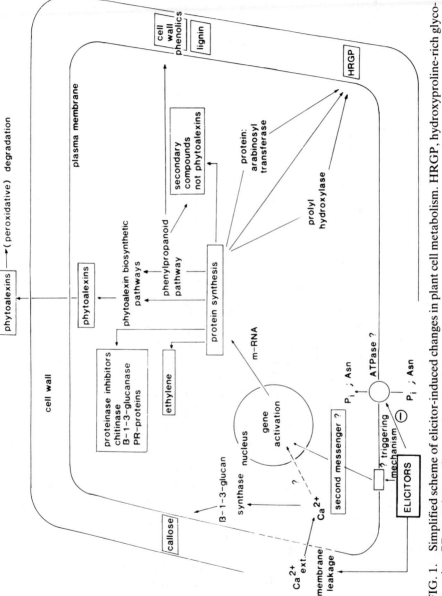

FIG. 1. Simplified scheme of elicitor-induced changes in plant cell metabolism. HRGP, hydroxyproline-rich glyco-proteins; PR-proteins, pathogenesis-related proteins.

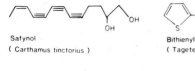
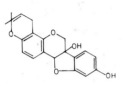
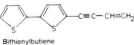
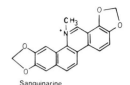
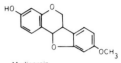

Safynol
(Carthamus tinctorius)

Bithienylbutiene
(Tagetes spec.)

Sanguinarine
(Papaver somniferum)

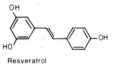
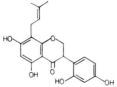
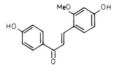

Glyceollin I
(Glycine max)

Kievitone
(Phaseolus vulgaris)

Medicarpin
(Cicer arietinum)

Resveratrol
(Arachis hypogaea)

Echinatin
(Glycyrrhiza echinata)

6-Methoxymellein
(Daucus carota)

Rishitin
(Solanum tuberosum)

Capsidiol
(Nicotiana tabacum)

Casbene
(Ricinus communis)

Psoralen
(Petroselium hortense)

Xanthotoxin
(Petroselium hortense)

Anthraquinones (Alizarin)
(Cinchona ledgeriana)

FIG. 2.   Structural diversity of phytoalexins isolated from plant cell suspension cultures.

these known plant defence compounds is steadily increasing (Bailey & Mansfield 1982, Darvill & Albersheim 1984, Dixon 1986). Features of the phytoalexin response in plant cell cultures are the rapidity of biosynthesis, the ability to be induced by defined chemical signals and the involvement of specific host genes which are only expressed under conditions of infection or cellular challenge. Apart from their biological function as defence compounds, phytoalexins are of interest because of their pronounced antimicrobial and possibly other pharmacological properties. Therefore, this area of natural product biochemistry may also be important for plant biotechnology and those interested in commercially useful plant products.

To illustrate the great diversity of chemical structures of phytoalexins, Fig. 2 presents selected examples of compounds isolated from various plant cell cultures. The structures include polyacetylenes, thiophenes, sesquiterpenes, diterpenes, coumarins, isocoumarins, chalcones, isoflavonoids, pterocarpans, stilbenes and alkaloids of different classes (DiCosmo & Misawa 1985). In several cases, a particular plant cell culture may produce a small number of closely related phytoalexins.

Investigations on isoflavanone, coumarine, pterocarpan and stilbene phytoalexins in plant cell cultures have demonstrated rapid, coordinate increases in the activity of sets of biosynthetic enzymes (Ryder et al 1986). Enzymes at branch points of pathways are especially prone to metabolic regulation. The gene systems concerned are highly polymorphic at the gene, mRNA and protein levels (Ebel 1986, Ryder et al 1986).

Isoflavone and phytoalexin metabolism in chickpea cell suspension cultures

The chickpea (*C. arietinum* L.) is an important crop plant in semiarid areas of the world. The most widespread fungal disease of chickpea is caused by the deuteromycete *Ascochyta rabiei*, which leads to the characteristic symptoms of 'Ascochyta blight'. Breeding programmes for *A. rabiei*-resistant cultivars (Saxena & Singh 1984) resulted in a great variety of differentially sensitive cultivars but there was no detailed biochemical investigation on antimicrobial defence reactions of chickpea plants and their cultivar-specific expression.

Extensive studies in the senior author's laboratory on *A. rabiei*-resistant (ILC 3279) and sensitive (ILC 1929) chickpea plants, as well as on various strains of the fungal parasite, have been started in order to describe the plant defence mechanisms and the properties of the fungus which determine virulence. Elucidation of initial biochemical events during host-parasite interaction has been greatly facilitated by using chickpea cell suspension cultures of the various cultivars and an *A. rabiei* polysaccharide elicitor (Kessmann et al 1987, Kessmann & Barz 1987, Barz et al 1987).

Our present knowledge of biochemical events during the interaction of chickpea with *A. rabiei* is summarized in Fig. 3. The main features of this interaction are:

1) degradation of plant cell wall polymers by fungal enzymes (Pandey et al 1987).
2) Liberation of an endogenous elicitor by plant enzymes or by wounding (Kessman & Barz 1986).
3) Occurrence of fungal inhibitors in the plant as pre-infectional isoflavones and post-infectional phytoalexins, with the latter being induced by both

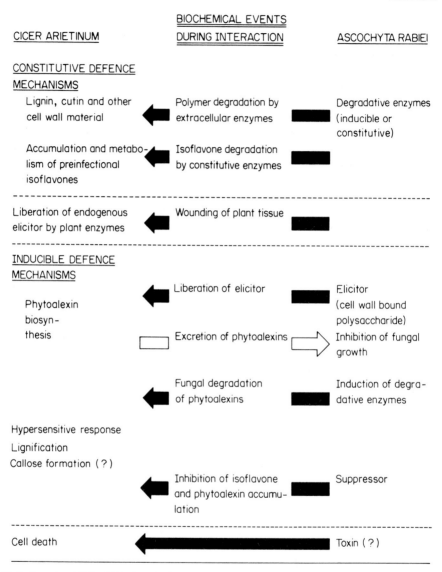

FIG. 3.   Established biochemical events during the interaction of the deuteromycete, *Ascochyta rabiei*, with its host plant, *Cicer arietinum*.

endogenous and fungal elicitors (Weigand et al 1986, Kessmann & Barz 1986, Barz et al 1987).

4)   The pronounced ability of virulent *A. rabiei* strains to degrade the isofla-

vones and the pterocarpan phytoalexins (Höhl & Barz 1987).

5) Inhibition of the accumulation of isoflavones, isoflavone conjugates and pterocarpan phytoalexins by a fungal suppressor (Kessmann & Barz 1986).

Fig. 3 also shows the complexity of this host–parasite interaction which requires much future investigation for complete understanding.

The importance of the 6aR:11aR-pterocarpans, medicarpin and maackiain, as the most prominent, rapidly accumulating, infection-induced antifungal compounds in a resistant but not in a susceptible cultivar of chickpea has been shown (Kessmann et al 1987, Weigand et al 1986). These investigations also demonstrated that the constitutively formed isoflavones, biochanin A and formononetin, together with their glucoside and malonylglucoside conjugates (Fig. 4), occur in these different cultivars at identical levels. Though a high polyphenol and isoflavone content of chickpea contributes significantly to the general resistance of this plant (Saxena & Singh 1984), these compounds should not be considered as a source of differential protection towards *A. rabiei*. The pronounced ability of various phytopathogenic fungi to degrade the defence compounds of the host, such as the dissimilation of the pterocarpan phytoalexins by *A. rabiei* (Höhl & Barz 1987), has been interpreted as a possibly important trait of fungal pathogenicity (VanEtten & Kistler 1984).

The aforementioned isolation of a preparation with suppressor activity from *A. rabiei* culture filtrate (Kessmann & Barz 1986) or germinating spores (H. Kessmann & W. Barz, unpublished work 1987) will be of special importance for our analyses of chickpea–*A. rabiei* interaction. Fungal suppressors are supposed to convert an incompatible interaction into a compatible one by eliminating or suppressing the host's defence mechanisms (de Wit 1986, Wood 1986). Furthermore, it has been suggested that the specificity of race–cultivar interaction may partly be explained by the existence of specific suppressor molecules. Our present investigations deal with the structural elucidation of the suppressor and biochemical measurements of the elicitor-suppressor antagonism. The preparation from *A. rabiei* (M_r 10 000–25 000; polysaccharide/glycoprotein; high affinity for concanavalin A; biological activity associated with the sugar moiety of the molecule) readily inhibits the accumulation of both the phytoalexins and isoflavones with their conjugates when tested in sliced cotyledons of chickpea (Kessmann & Barz 1986). The mode of action of suppressors, their exact role in pathogenicity and the mechanism of eliminating the effects of elicitors remain to be elucidated.

During our investigations on the biosynthesis of the constitutively expressed pre-infectional and the inducible postinfectional inhibitors of *C. arietinum*, the enzymology of the formation and subsequent metabolism of the isoflavone malonylglucosides has been clarified (Barz et al 1987). Starting with the isoflavone aglyca, the metabolic grid (Fig. 4) involves 1) an isoflavone-specific UDP-glucose: 7-O-glucosyltransferase; 2) a malonyl-CoA-dependent

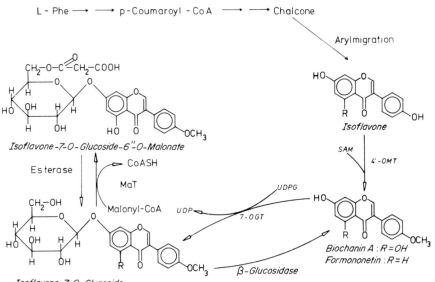

FIG. 4. Metabolic grid of an enzyme sequence showing the conjugation reactions and the metabolic turnover of the isoflavones in *Cicer arietinum*. Abbreviated enzymes are: 4'-OMT:S-Adenosylmethionine, isoflavone 4'-O-methyltransferase; 7-OGT:UDPG, isoflavone 7-O-glucosyltransferase; MAT: Malonyl-CoA, isoflavone 7-O-glucoside-6"-O-malonyltransferase. (Review: Barz et al 1987.)

malonyl transferase with specificity for carbon 6 of glucose in isoflavone 7-O-glucosides; 3) a highly substrate-specific malonyl esterase; and 4) several isozymes of isoflavone glucoside-specific glucohydrolases. These enzymes of isoflavone metabolism (Hinderer et al 1986) have been found both in plants and in chickpea cell suspension cultures producing isoflavone conjugates. This metabolic grid also explains the established turnover of formononetin conjugates and the proposed metabolic link between the metabolism of isoflavone conjugates and the biosynthesis of pterocarpan phytoalexins. In this linkage of two related pathways, formononetin functions as the central intermediate (see Fig. 5).

Accumulation of medicarpin and maackiain in chickpea cell suspension cultures represents an elicitor (yeast extract; *A. rabiei* polysaccharide) induced process involving *de novo* enzyme synthesis, as shown by the action of molecular inhibitors of transcription and translation (Barz et al 1987). The response is quite rapid; maximum levels of the phytoalexins are reached some eight hours after elicitor application. A most remarkable aspect of using cell suspension cultures of cultivar ILC 3279 (resistant; high phytoalexin production) and ILC 1929 (susceptible; 3–4 times lower production of phytoalexins with medicarpin as the only product) is that the cell cultures still show the same highly differential pterocarpan formation as the intact plants (Kessmann

& Barz 1987, Barz et al 1987). Such cultures can therefore be used for biochemical and molecular biological studies on phytoalexin biosynthesis, regulation of the expression of enzymes for defence reactions and their pathways in the two chickpea genotypes. On the other hand, cell suspensions of cultivars ILC 1929 and ILC 3279 show no difference in their concentrations of isoflavone and isoflavone conjugates (Kessmann & Barz 1987).

The pterocarpan phytoalexins formed by cultured chickpea cells are mainly excreted into the growth medium rather than being deposited in the cell vacuole. Similar observations have been made for other elicitor-induced compounds (coumarins, thiophenes, alkaloids). The mechanism of this process is still completely unknown. After excretion, the pterocarpans are rapidly polymerized or degraded by extracellular peroxidases (U. Jaques et al, unpublished work 1987). Comparative analyses of pterocarpan peroxidative destruction in chickpea cultures ILC 1929 and ILC 3279 have shown that these side reactions occur in both cultures to the same extent, so that the much lower productivity of pterocarpans in culture ILC 1929 cannot be explained by preferential peroxidative dissimilation.

Treatment of chickpea cell suspension culture ILC 3279 with elicitors allowed the isolation and characterization of three new enzymes involved in late stages of pterocarpan biosynthesis. The substrate specificities of the microsomal formononetin 2'-hydroxylase and the 3'-hydroxylase (Hinderer et al 1987), together with the NADPH:2'-hydroxyisoflavone oxidoreductase (Tiemann et al 1987), indicate the sequence of intermediates in the pterocarpan-specific biosynthetic pathway shown in Fig. 5. These results confirm earlier data on pterocarpan formation obtained by feeding experiments with labelled precursors (Dixon 1986). The same intermediates could also be isolated from among a multitude of newly accumulating polyphenols in elicitor-treated, sliced cotyledons of chickpea (Jaques et al 1987). This experimental approach of applying elicitors to sliced tissues should be very useful for the detection of new biologically active compounds from plants. The three new enzymes of pterocarpan biosynthesis (Fig. 5) belong to the comparatively small group of characterized enzymes specifically involved in phytoalexin biosynthesis (Dixon 1986, Barz et al 1987), which may be used for future detailed investigations on the gene expression of elicitor-induced plant defence reactions.

Fungal elicitors cause substantial alterations in cellular metabolism (Fig. 1). These observations have led to extensive studies in chickpea cell culture ILC 3279 by measuring elicitor effects on a total of 15 enzymes of primary and secondary metabolism (S. Daniel and W. Barz, unpublished work 1987, U. Flentje and W. Barz, unpublished data 1987, Barz et al 1987). Among the enzymes of primary metabolism, only glucose-6-phosphate dehydrogenase showed a pronounced transient increase in the specific activity. The same response was obtained for the constitutively formed enzymes of the general

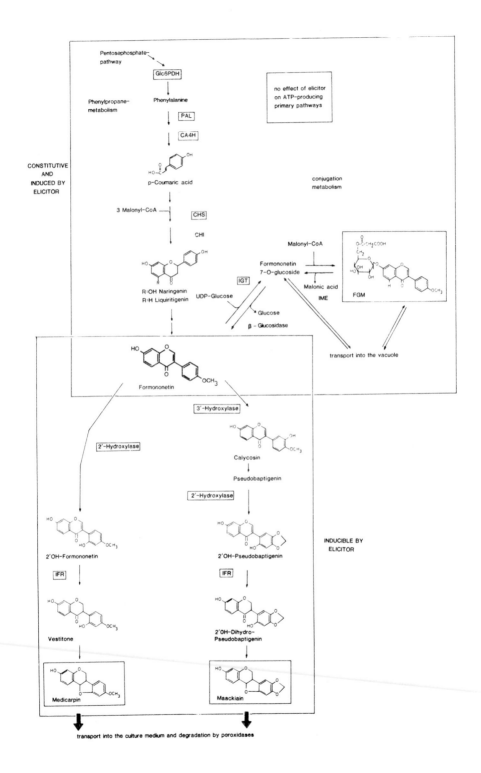

Pentosephosphate-
pathway

Glc6PDH

Phenylpropane-
metabolism

Phenylalanine

PAL

CA4H

no effect of elicitor
on ATP-producing
primary pathways

CONSTITUTIVE
AND
INDUCED BY
ELICITOR

p-Coumaric acid

conjugation
metabolism

3 Malonyl-CoA

CHS

CHI

Malonyl-CoA

Formononetin
7-O-glucoside

Malonic acid

FGM

R=OH Naringenin
R=H Liquiritigenin

UDP-Glucose

IGT

IME

Glucose

β - Glucosidase

transport into the vacuole

Formononetin

3'-Hydroxylase

Calycosin

2'-Hydroxylase

Pseudobaptigenin

2'-Hydroxylase

INDUCIBLE BY
ELICITOR

2'OH-Formononetin

2'OH-Pseudobaptigenin

IFR

IFR

Vestitone

2'OH-Dihydro-
Pseudobaptigenin

Medicarpin

Maackiain

transport into the culture medium and degradation by peroxidases phenylpropanoid pathway (L-phenylalanine ammonia lyase, cinnamic acid-4-hydroxylase, chalcone synthase) and of the isoflavone conjugation reactions (isoflavone-specific 7-O-glucosyl-transferase). The elicitor-induced increase in the enzyme activity of this glucosyltransferase (IGT in Fig. 5) must be seen in connection with the elicitor-dependent increased accumulation of the formononetin 7-O-glucoside-6″-O-malonate (FGM in Fig. 5). All other isoflavones, such as biochanin A, formononetin and biochanin A 7-O-glucoside-6″-O-malonate, showed no elicitor-dependent changes in their concentrations. These and other investigations, together with results on the elicitor induction of the pterocarpan-specific biosynthetic enzymes are summarized in Fig. 5. Three sets of enzymes (phenylpropane metabolism, isoflavone conjugation reactions and the phytoalexin-specific branch) with formononetin as the central intermediate appear to be metabolically linked in elicitor-caused alterations of enzyme activities. Enzymes at branch points of pathways seem to be especially affected in such elicitor-caused changes of metabolism. The chickpea cell culture system established so far will undoubtedly allow detailed further investigations on such important questions as the involvement of elicitor-induced isoenzymes, the mechanisms of the transient increase of enzyme activities and the number and activation processes of the genes involved.

Preliminary data obtained from comparative enzyme measurements in elicitor-treated chickpea cell suspension cultures ILC 3279 and ILC 1929 have shown that L-phenylalanine ammonia lyase activity is expressed at identical levels, whereas the isoflavone 2′-hydroxylase and the NADPH:isoflavone oxidoreductase reach only very low levels in the cell culture established from the susceptible genotype (Kessmann et al 1987).

Conclusions

The chickpea cell suspension cultures established from *A. rabiei*-susceptible and resistant genotypes provide elegant systems for elucidating biochemical

---

FIG. 5. Postulated scheme of three sets of metabolically linked pathways (phenylpropane metabolism, isoflavone conjugation reactions, pterocarpan-specific biosynthetic branch) in elicitor-treated chickpea cell suspension cultures ILC 3279 with formononetin as the central intermediate. Enzymes and compounds shown in frames are subject to elicitor-caused increases in specific activity or *de novo* synthesis, and higher accumulation, respectively. The unframed enzymes and compounds indicate that no elicitor effect could be measured. Abbreviated enzymes are: Glc6PDH, glucose-6-phosphate dehydrogenase; PAL, L-phenylalanine ammonia lyase; CA4H, cinnamic acid 4-hydroxylase; CHS, naringenin-chalcone synthase; CHI, chalcone isomerase; IGT, isoflavone-specific UDP-glucose:7-O-glucosyltransferase; IME, isoflavone malonylglucoside-specific: malonylesterase; 2′-hydroxylase, isoflavone 2′-hydroxylase; 3′-hydroxylase, isoflavone 3′-hydroxylase; IFR:NADPH, isoflavone oxidoreductase;FGM, formononetin 7-O-glucoside-6″-O-malonate.

events in host–parasite interaction and elicitor-caused alterations of cellular metabolism. Results obtained hitherto show that these cultures will allow comparative investigations on the mode of differential plant resistance and on the metabolic regulation and expression of the phytoalexin response, which represents one of the most essential antimicrobial defence reactions of higher plants.

Acknowledgements

Financial support by Deutsche Forschungsgemeinschaft and Fonds der Chemischen Industrie is gratefully acknowledged.

References

Bailey JA, Mansfield JA 1982 Phytoalexins. Blackie, Glasgow & London

Barz W, Daniel S, Hinderer W et al 1987 Elicitation and metabolism of phytoalexins in plant cell cultures. In: Pais MS (ed) Plant cell biotechnology. Springer-Verlag, Berlin, Heidelberg, in press

Darvill AG, Albersheim P 1984 Phytoalexins and their elicitors. A defence against microbial infection in plants. Annu Rev Plant Physiol 35:243–276

de Wit PJGM 1986 Elicitation of active resistance mechanisms. In: Bailey JA (ed) Biology and molecular biology of plant-pathogen interactions. Springer-Verlag, Berlin, Heidelberg, New York, p 149–169

DiCosmo F, Misawa M 1985 Eliciting secondary metabolism in plant cell cultures. Trends Biotechnol 3:318–322

Dixon RA 1980 Plant tissue culture methods in the study of phytoalexin induction. In: Ingram DS, Helgeson JP (eds) Tissue culture methods for plant pathologists. Blackwell, Oxford, p 185–196

Dixon RA 1986 The phytoalexin response: elicitation, signalling and control of host gene expression. Biol Rev 61:239–291

Ebel J 1986 Phytoalexin synthesis: the biochemical analysis of the induction process. Annu Rev Phytopathol 24:235–264

Hinderer W, Köster J, Barz W 1986 Purification and properties of a specific isoflavone 7-0-glucoside-6″-malonate malonylesterase from roots of chickpea (*Cicer arietinum* L.). Arch Biochem Biophys 248:570–578

Hinderer W, Flentje U, Barz W 1987 Microsomal isoflavone 2′-and 3′-hydroxylases from chickpea (*Cicer arietinum* L.) cell suspension cultures induced for pterocarpan phytoalexin formation. FEBS (Fed Eur Biochem Soc) Lett 214:101–106

Höhl B, Barz W 1987 Partial characterization of an enzyme from the fungus *Ascochyta rabiei* for the reductive cleavage of pterocarpan phytoalexins to 2′-hydroxyisoflavans. Z Naturforsch Sect C Biosci 42:897–901

Jaques U, Kessmann H, Barz W 1987 Accumulation of phenolic compounds and phytoalexins in sliced and elicitor-treated cotyledons of *Cicer arietinum* L. Z Naturforsch Sect C Biosci 42:1171–1178

Kauss H 1987 Some aspects of calcium-dependent regulation in plant metabolism. Annu Rev Plant Physiol 38:47–72

Kessmann H, Barz W 1986 Elicitation and suppression of phytoalexins and isoflavone accumulation in cotyledons of *Cicer arietinum* L. as caused by wounding and by polymeric components from the fungus *Ascochyta rabiei*. J Phytopathol 117:321–335

Kessmann H, Barz W 1987 Accumulation of isoflavones and pterocarpan phytoalexins in cell suspension cultures of different cultivars of chickpea (*Cicer arietinum*). Plant Cell Rep 6:55–59

Kessmann H, Tiemann K, Jansen JR, Reuscher H, Bringmann G, Barz W 1987 In vivo characterization of NADPH: 2'-hydroxyisoflavone oxidoreductase in elicitor treated chickpea cell cultures and stereochemical aspects of the phytoalexins medicarpin and maackiain. In: Pais MS (ed) Plant cell biotechnology. Springer-Verlag, Berlin, Heidelberg, in press.

Kombrink E, Bollmann J, Hauffe KD et al 1986 Biochemical responses of non-host plant cells to fungi and fungal elicitors. In: Bailey JA (ed) Biology and molecular biology of plant-pathogen interactions. Springer-Verlag, Berlin, Heidelberg, New York, p 253–262

Kurosaki F, Tsurusawa Y, Nishi A 1987 The elicitation of phytoalexins by Ca^{2+} and cyclic AMP in carrot cells. Phytochemistry (Oxf) 26:1919–1923

Pandey BK, Singh US, Chaube HS 1987 Mode of infection of Ascochyta blight of chickpea caused by *Ascochyta rabiei*. J Phytopathol 119:88–93

Ryder TB, Bell JN, Cramer CL et al 1986 Organization, structure and activation of plant defence genes. In: Bailey JA (ed) Biology and molecular biology of plant-pathogen interactions. Springer-Verlag, Berlin, Heidelberg, New York, p 207–219

Saxena MC, Singh KB 1984 Ascochyta blight and winter sowing of chickpea. Kluwer, The Hague, The Netherlands & Boston & Lancaster

Tiemann K, Hinderer W, Barz W 1987 Isolation of NADPH:isoflavone oxidoreductase, a new enzyme of pterocarpan phytoalexin biosynthesis in cell suspension cultures of *Cicer arietinum*. FEBS (Fed Eur Biochem Soc) Lett 213:324–328

VanEtten HD, Kistler HC 1984 Microbial enzyme regulation and its importance for pathogenicity. In: Kosugue T, Nester EW (eds) Plant-microbe interactions. Molecular and genetic perspectives. Macmillan, New York, vol 1:42–68

Ward EWB 1986 Biochemical mechanisms involved in resistance of plants to fungi. In: Bailey JA (ed) Biology and molecular biology of plant-pathogen interactions. Springer-Verlag, Berlin, Heidelberg, New York, p 107–131

Weigand F, Köster J, Weltzien HC, Barz W 1986 Accumulation of phytoalexins and isoflavone glucosides in a resistant and a susceptible cultivar of *Cicer arietinum* during infection with *Ascochyta rabiei*. J. Phytopathol 115:214–221

Wood RKS 1986 Introductory comments on host-parasite interaction. In Bailey JA (ed) Biology and molecular biology of plant-pathogen interactions. Springer-Verlag, Berlin, Heidelberg, New York, p 1–13

DISCUSSION

Galun: One thing is quite interesting from a phytopathological point of view. If I understood you correctly, the defence mechanism of resistant plants is achieved by a higher or more efficient production of phytoalexins after the elicitors are given to the plant or the plant cells. Many commercial companies have been interested in making artificial elicitors to non-resistant varieties. It seems that you must have resistant varieties before you can induce a better defence mechanism, you cannot improve the defence mechanism of susceptible varieties. Is that true?

Barz: What you describe would be a kind of vaccination of the plant with natural or with artificial elicitors.

Galun: Provided the variety is resistant as such.

Barz: Yes, although I think it would not be desirable to have high levels of phytoalexins present in the plant at all times because these compounds are undoubtedly toxic to cells. The essential point is that the plant has the ability to form these compounds very rapidly, so I think one should somehow alter the sensitivity of the triggering mechanism—the faster these compounds are formed around the site of infection, the better. Analyses of the time intervals required for the onset of phytoalexin biosynthesis after elicitor application have shown that the lag phase may be as short as five or ten minutes. After such a short period one can already detect mRNA synthesis. Therefore, the more rapid phytoalexin accumulation occurs, the better.

Galun: But the cultivar is always important?

Barz: The cultivar is important. There is another point which warrants detailed investigation. A broad screening programme involving a great number of cultivars should be initiated to try to correlate resistance towards a specific race of a fungal pathogen with the quantitative level of phytoalexin production. Although there can be no doubt that phytoalexins are essential, a quantitative correlation has not yet been measured.

Harms: Do you have any suspicion where in the susceptible variety the lesion is that causes this inability to respond? Is there accumulation of an intermediate product which would indicate that a specific enzyme in the biosynthetic pathway is non-functional?

Is my impression correct that the purification of some of these enzymes has progressed to a state where they are pure enough to be used for raising monoclonal antibodies and for sequencing? I got the impression that the time is ripe for the molecular cloning of some of those enzymes and for some exciting immunochemistry and molecular analysis that could elucidate the subcellular localization of the pathways and the transport systems.

Barz: With regard to the enzymes involved in pterocarpan biosynthesis, the answer is yes, except for the isoflavone mono-oxygenases which are bound to microsomes. For the other soluble enzymes, antibodies are currently being raised.

Potrykus: From what I have heard about these elicitors and phytoalexins, I am worried about the lack of specificity of the responses, which contrasts with the high specificity of the response of the plant to pathogens. Is it possible that one is studying unspecific side effects at the moment and is missing the important specific effects?

Barz: Nobody has ever claimed that the known effects are the only ones which occur. There might be some other effects and these may be far more specific. The work which I refer to with the phytoalexins, especially the biosynthetic work, has been performed with numerous cell cultures; they are model systems, with all the advantages but also with some limitations. The cell culture can never totally simulate the situation in an intact plant. On the other hand, there can be no doubt that the phytoalexin response is rather rapid and that it belongs to the initial responses of the plant cell. As we develop more refined analytical techniques, for example with antibodies, we see that the response occurs within 20 to 60 minutes after the signal has been received. There are several reports which clearly show that the phytoalexins are essential for establishing resistance. The present data cannot be used to explain resistance completely, there are other factors which also contribute to what you finally see as resistance.

Riley: But does the distinction between the two chickpea cultivars that you described owe its origin to a distinction at a single genetic locus?

Barz: Yes, as far as the plant breeders and geneticists have analysed it, this is a monogenic trait inherited semi-dominantly.

Riley: And so the effect of that locus brings into play quite different pathways?

Barz: Exactly, that's just what I wanted to mention: we should not take this single locus as affecting only one enzyme. It could be a regulatory switch, so that a whole sequence of events is governed by this locus.

Potrykus: This illustrates my worry that one is engaged in studying branch ends of this reaction, and somehow loses the focus of the starting point of these reactions. In this case, if this is a monogenic trait, then somewhere this starts with probably one product, possibly one enzyme. My general impression from listening to speakers at conferences (I am not reading in this field) is that the analysis goes into great depth chemically somewhere at the periphery but one is losing touch with the basic issue.

Barz: What do you mean by basic? The fact is that the resistance genes have been known for a number of years for several plant species, but so far, with possibly one or two exceptions, nobody really knows what the product of any of those genes is. I have the impression that we should at least contribute to the understanding of this field by performing biochemical analyses to see what happens in such interactions. That's why we have started by asking the question, what is the biochemical difference between these two cultivars? We have chosen extreme cultivars, highly resistant and highly sensitive. There is quite a long list of factors which in my personal opinion should be analysed in our search for the cause of determination of resistance against fungal pathogens. You can transfer resistance by various techniques which we have heard in the last two days but this does not give you any biochemical or chemical explanation of what actually happens. How essential the various factors are, how they contribute to virulence or resistance, the future has to show. We have to take these complex plant-parasite interactions apart and analyse them. From this point of view, molecular biology offers the ideal techniques, because we can pick out various genes and express them in heterologous systems. We are currently working with the inducible enzymes and their genes involved in the fungal degradation of the phytoalexins, in order to determine precisely what is their contribution to fungal virulence.

Riley: The important thing is that substitution of the allele giving resistance for the allele giving susceptibility by backcrossing into the same genetic background causes resistance. Therefore the pathway to resistance must have been present but not induced; the distinction is that in some cases the pathway is brought into play and in others it isn't. If that is the case, it's very important for genetic engineering, because you can ignore the rest of the pathway and simply engineer the triggering gene.

Barz: That is correct and programmes for this are being initiated in collaboration with ICARDA at Alepppo, Syria.

Hall: I suspect Professor Barz is more familiar than I with the work of Dr Fritig at Strasbourg. They are looking at some of these general detoxification or pathogen responses where many plants have a common set of pathogenesis-related proteins. Dr Fritig has characterized quite a large number of these pathogenesis-related proteins, which have proved to be chitinases, namely N-acetylglucosamine glycanohydrolase and β-1,3 glucanases (Legrand et al 1987). These proteins clearly relate to some of the reactions that you are talking about. They are self-stimulating in that once the fungal elicitor activates the plant enzymes, they even degrade your 'endoelicitors' to make the plant reaction self-perpetuating.

Potrykus: But the worrying thing is again that you can induce these chitinases by treatment with viruses, which doesn't make too much biological sense, because I do not think that viruses contain chitin.

Hall: I disagree; plants clearly do not have specific immune systems in the way that animals have, but they do have generalized systems which basically cut off the accessibility of the attacked cells. This also relates to phenomena such as suberization. So it doesn't really matter at one level whether the agressor happens to be a virus or an insect or a fungal pathogen, the general reaction of a cell under attack is to close itself off from the rest of the plant.

Tabata: It is very interesting that in chickpea cell cultures the phytoalexins produced are secreted or excreted into the medium, instead of being accumulated in vacuoles. Do you have any explanation for this peculiar phenomenon? This may have an interesting application for inducing excretion of secondary metabolites from cultured cells.

Barz: This excretion of phytoalexins is by no means confined to the chickpea; it has also been observed in a number of other cases. Personally, I am almost inclined to redefine the term phytoalexin as 'a compound which after *de novo* synthesis is excreted from the cell'. There is an overwhelming body of information that they are excreted. From a biological point of view it makes sense because the compound should not act in the synthesizing cell but outside, where the enemy is. If we could characterize the mechanisms of excretion, it would be very helpful to us so that we can re-direct cellular transport from the vacuole to the outside of the cell.

Tabata: Do you think that these flavonoids could be accumulated in vacuoles but in your case they are not.

Barz: Many of the phytoalexins have hydroxyl groups but they are not glucosylated or glycosylated. With the chickpea cell culture the difference is quite clear: after addition of an elicitor both an isoflavone malonyl-glucoside and two phytoalexins are made at increased concentrations. The isoflavone conjugate is deposited in the vacuole, whereas the two phytoalexins are excreted into the nutrient medium.

Tabata: So you suppose that there may be an important change in the permeability or structure of the plasma membrane caused by the elicitor?

Barz: It's quite possible.

Zenk: I am not really convinced that excretion of secondary compounds is a normal pathway. You may be right for some of the phytoalexins but I know of only one single case, that of Professor Tabata's group, which is berberine secretion by the Thalictrum minus system. They have shown in a precise way that the secondary compound is excreted into the medium. I think under non-elicitation conditions there are extremely few truly excreting cell lines.

Barz: That's exactly what I mean, the normal cellular excretion of a plant cell is into the vacuole. But in the case of phytoalexins, regardless of how we define what phytoalexins are, they are synthesized *de novo* after application of elicitor, and subsequently a great number, almost 100%, are totally or partially excreted from the cells.

Zenk: I can give you at least one system where this is definitely not true. This is our *Eschscholtzia* system where we can elicit benzophenanthridine synthesis with a yeast elicitor (Schumacher et al 1987). These benzophenanthridines, as far as we know from Kohlenbachs' work, are in the vacuole (Lang & Kohlenbach 1982).

Barz: Let me give you another example to stress my point. In the case of the alkaloid sanguinarine from *Papaver somniferum* cell cultures, W. Kurz and co-workers, in Sastoon, Canada, have convincingly demonstrated that this elicitor-induced compound is excreted from the cells into the medium. It has been shown that the cells can be removed by filtration leaving the bulk of the alkaloid in the medium.

Zenk: I think it's very similar to the situation where one has a certain amount of dying cells which release the products into the medium.

Barz: Excretion of phytoalexins cannot be explained by cell lysis.

Fowler: I would like to support the previous comment. We have a number of cell lines in our culture bank which release large amounts of material, as much as 85% of the total culture content being a particular compound. The viability of these cultures measured not by fluoroscein diacetate but by growing on, a true viability test, is well above 90%. To balance this statement, I must say that we also have cells where very little material is released into the external medium.

Potrykus: If I am not mistaken, chitinase fits the criteria for a classical phytoalexin — it's elicitor induced, it is supposed to act against phytopathogens including even insects.

Barz: You call chitinase a phytoalexin?

Potrykus: What would you call it?

Barz: A pathogen-induced protein. It is clearly directed against the microorganism but it does not fulfil the requirement that it is of low molecular weight.

Potrykus: It is not secreted, as far as I know it is secreted into the vacuole.

Barz: No, several chitinases are excreted.

Tabata: When we screened various plants for antidermatophytic substances, we isolated plenty of maackiain from the root of *Sophora angustifolia* as a main antidermatophytic substance. As you know Professor Furuya (1968) also has isolated a lot of maackiain from callus cultures of *S. angustifolia*. Would you say that such a compound could be called a phytoalexin? Are there any phytoalexins known which have never been found in normal plants?

Barz: Maackiain is a phytoalexin, not an elicitor. I think that these compounds may be called phytoalexins, although maackiain and some of the other pterocarpans occur naturally in the hardwood of several tropical trees. They contribute to the resistance of this wood against microbial decay. But otherwise, in the herbaceous plant, they are formed in response to elicitation.

Fowler: To move to cell suspensions: when you apply the elicitor to the cell suspension, is it a continuous supply or a single dose? If you look at the increase in enzyme activities, of which you showed a wide range, after you supply the elicitor, how long does the enzyme activity remain elevated? Have you any indication of the half-life of those enzyme systems?

Barz: We add, as most people do, the elicitors in batches approximately every 24 hours. Thus, we observe a re-elicitation step about every 24 hours with subsequent phytoalexin formation after the addition of each batch of elicitor.

Fowler: After that second elicitation, is there an absolute increase in enzyme activity or does it simply return to the original high level?

Barz: It returns to the original level; this is a transient accumulation of the enzymes.

Fowler: So it isn't additive, you actually reach a maximum capacity?

Barz: By using combinations of elicitors, one can obtain an additive effect on the level of enzyme but of course that cannot be carried on indefinitely.

Durzan: The results are interesting and they suggest that there might be plant survival in a culture situation after exposure to an invading organism. After the elicitation and defence events have occurred, is there anything amongst the properties of the induced chemicals that will allow the surviving cells in the culture to recover and return to normal development?

Barz: They grow happily without symptoms of toxication; they are not killed, is that what you mean?

Durzan: I mean that there is always a growth or developmental response in culture situations. There have been reports that compounds of this type might form conjugates with nitrogenous compounds and these compounds may actually promote development. It's one thing to survive invasion by a pathogen, but after the invasion and after the survival, are there any properties among the new products that help the cell promote and programme future activities, in a developmental sense?

Barz: The cell culture system is very different from the situation in the intact plant because in the cell culture the compounds are not only excreted into the culture medium but they are also peroxidatively polymerized and degraded. Of the fifteen compounds that I showed on one of my slides, about three-quarters are sensitive to peroxidative destruction. Berberine is not peroxidatively attacked and thus we observe an accumulative effect. In such a case, the culture has to deal with this material by directing it into alternative reactions.

Durzan: What I am trying to get at is that in many embryo regeneration situations, there is initially a lot of stress or environmental problems and many of the cells may deteriorate. Often one sees accompanying browning reactions, a lot of fluorescent compounds appearing, and then one sees an embryo appearing amongst this black cell mass. The question is, are there any new chemical properties in that environment or reactions that enhance somatic embryogenesis?

Barz: This is the first time that I have heard that.

Zenk: I think this phytoalexin concept is very attractive for breeders in the future and also for breeders who are engaged in optimizing food production. On the other hand, these phytoalexins sometimes accumulate in very high amounts, up to 1%, 2%, 3% (dry weight) in a given plant. Do you see any toxicological problems? Do you think that, especially if breeders are taking exotic genes and transferring them into food plants, legislators may move in and say that you have to determine the toxicology of these compounds?

Barz: Very little work has been done on the toxicology of these compounds in humans. Looking at some of the structures, I would prefer to have them checked from a pharmacological or toxicological point of view before I would eat plant material containing high amounts of phytoalexins. There is an interesting example in mammals. A few weeks ago we were approached by scientists from Australia who have observed a relevant situation in sheep which have been fed on clover. Clover plants contain comparatively low levels of isoflavones but these compounds are highly oestrogenic. If the clover fields are either sprayed with certain chemicals, such as a fungicide, or infected with fungal parasites, there is a general increase within the clover leaves of these highly oestrogenic compounds. This appears to be a clear elicitation phenomenon. As a result, disturbances of the reproductive cycle of the sheep were observed. However, I am sure that at some point we have all eaten phytoalexins.

References

Furuya T, Ikuta A 1968 The presence of *l*–maackiain and pterocarpin in callus tissue of *Sophora angustifolia*. Chem Pharm Bull 16:771

Honda G, Tabata M 1982 Antidermatophytic substances from *Sophora angustifolia*. Planta Med *1982*:122-123

Lang H, Kohlenbach HW 1982 Differentiation of alkaloid cells in cultures of macleaya mesophyll protoplasts. Planta Med 46:78–81

Schumacher HM, Gundlach H, Fiedler F, Zenk MH 1987 Elicitation of benzophenan-thridine alkaloid synthesis in Eschscholtzia cell cultures. Plant Cell Rep 6:410–413

Biosynthesis of tropane alkaloids

Yasuyuki Yamada and Takashi Hashimoto

Research Center for Cell and Tissue Culture, Faculty of Agriculture, Kyoto University, Kyoto 606, Japan

Abstract. Roots of various solanaceous plants grow well *in vitro* and produce large amounts of tropane alkaloids. These root cultures provide excellent materials for tracer experiments and for the extraction of the alkaloid biosynthetic enzymes. Our enzymic and tracer studies on the early biosynthesis of tropane alkaloids indicated that N-methylputrescine is synthesized mostly from symmetrical putrescine in our root cultures; these results are contrary to the previous asymmetrical biosynthetic scheme.

We found an enzymic activity that hydroxylates hyoscyamine at the C-6 position in cell-free extracts from various root cultures. The enzyme hyoscyamine 6β-hydroxylase (EC 1.14.11.-) requires *l*-hyoscyamine, 2-oxoglutarate, ferrous ions and molecular oxygen for activity. Ascorbate and catalase stimulate the hydroxylation. The purified enzyme epoxidized 6,7-dehydrohyoscyamine, a hypothetical precursor of scopolamine, to scopolamine. Several lines of evidence indicate that a single 2-oxoglutarate-dependent dioxygenase catalyses both the hydroxylation and the epoxidation reactions.

Shoot cultures of alkaloid-producing plants are unique in that they do not accumulate alkaloids, yet they will convert added hyoscyamine to scopolamine. When [6-^{18}O]6β-hydroxyhyoscyamine was fed to *Duboisia* shoot cultures, the labelled alkaloid was converted to scopolamine which retained ^{18}O in the epoxide oxygen. Therefore, 6β-hydroxyhyoscyamine is converted *in vivo* to scopolamine without a dehydration step. Some other interesting reactions in the tropane alkaloid biosynthesis are also presented.

1988 Applications of plant cell and tissue culture. Wiley, Chichester (Ciba Foundation Symposium 137) p 199–212

The molecular structure of tropane alkaloids is characterized by an 8-azabicyclo[3.2.1]octane ring system. The N-methyl derivative is commonly called tropane, and many alkaloids are derivatives of tropine, i.e. tropane-3α-ol. Tropine derivatives esterified with tropic acid (e.g. hyoscyamine and scopolamine) are found in several *Solanaceae* species: *Atropa, Datura, Duboisia, Hyoscyamus* and *Scopolia* (Evans 1979, Romeike 1978). These tropine esters and their synthetic derivatives are important anticholinergic agents that are used for their antisecretory and antispasmodic effects on the gastrointestinal tract and their depressant effects on tremors, to name just a few.

On the basis of extensive tracer experiments, a biosynthetic pathway has

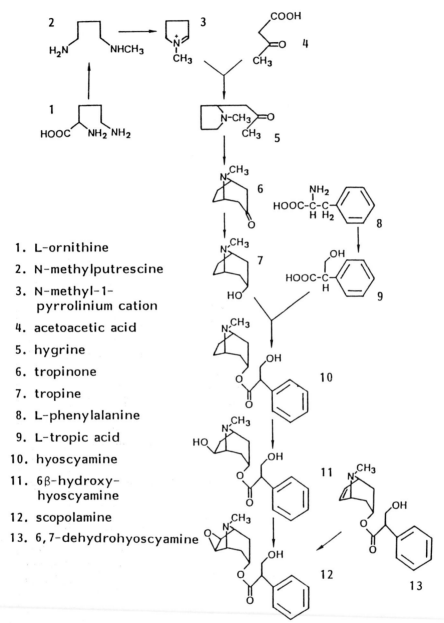

1. L-ornithine
2. N-methylputrescine
3. N-methyl-1-
   pyrrolinium cation
4. acetoacetic acid
5. hygrine
6. tropinone
7. tropine
8. L-phenylalanine
9. L-tropic acid
10. hyoscyamine
11. 6β-hydroxy-
    hyoscyamine
12. scopolamine
13. 6,7-dehydrohyoscyamine

FIG. 1.   Biosynthesis of tropane alkaloids. Each arrow does not necessarily indicate one reaction step.

now been proposed (Leete 1979). Fig. 1 outlines this proposed pathway, most of which has yet to be substantiated by the isolation of the component biosynthetic enzymes. In this paper, we summarize our recent discovery of

TABLE 1   Tropane alkaloid contents (% dry weight) in Hyoscyamus niger L

Source	Hyoscyamine	Scopolamine	Scop/Hyos
Intact leaves	0.01–0.04	0.01 –0.05	1–3
Intact mature roots	0.01–0.02	0.01 –0.04	1–3
Cultured cells	0.05–0.08	0.001–0.003	0.1–0.3
Cultured roots	0.04–0.08	0.12 –0.30	3–8

several of these enzymes. Their existence necessitates a revision of the previously proposed pathway. Our studies have been made possible by the use of tissue cultures of alkaloid-producing plants.

Plant tissue culture

Tropane alkaloids are not produced in large amounts by undifferentiated cells (Table 1). Although repeated screening for the most productive cell lines and optimization of culture conditions for greater alkaloid content resulted in suspension cultures of *Hyoscyamus niger* with hyoscyamine content higher than that of the intact plant, the scopolamine content of the culture remained one-tenth of the content of the intact plant (Yamada & Hashimoto 1982, Hashimoto & Yamada 1987b). For scopolamine biosynthesis, differentiated root tissue is preferable (Hashimoto & Yamada 1983). Root cultures of several solanaceous species were screened and optimal culture conditions were investigated for rapid growth and increased alkaloid formation (Hashimoto et al 1986). For example, we have selected root cultures of *H. niger* and *H. albus* which exhibit a fivefold fresh weight (FW) increase in one week and contain 0.3% dry weight (DW) of scopolamine (*H. niger*) and 0.6% DW of hyoscyamine (*H. albus*). Starting from cultured roots (2.5–3.0g FW) in one 100 ml flask, we are routinely harvesting one kilogram FW of cultured roots for enzyme extraction after one month. Transformation of our *Hyoscyamus* root cultures (*H. niger, H. albus* and *H. gyorffi*) with *Agrobacterium rhizogenes* strain 15834 did not markedly increase the growth rate or the alkaloid production in these cultures (unpublished data).

Shoot cultures of alkaloid-producing plants also provide unique materials for the study of tropane alkaloid biosynthesis. *Duboisia* shoot cultures, for example, contain no alkaloids, but if hyoscyamine is added to the culture medium it is converted to scopolamine (Table 2) (Griffin 1979, Hashimoto et al 1987). These shoot cultures must be incapable of synthesizing hyoscyamine or some earlier precursor of hyoscyamine, thus preventing accumulation of alkaloids. The presence of an active pathway from hyoscyamine to scopolamine in the absence of alkaloids in the cultures is especially useful for the feeding studies with precursors labelled with stable isotopes (Hashimoto et al 1987).

TABLE 2   Biotransformation of tropane alkaloids in Duboisia myoporoides shoot cultures

Alkaloid added (2µmol/flask)	Alkaloids detected after one week				Recovery
	Hyoscyamine	Hyos-OH	De-hyos	Scopolamine	
	µmol/flask				%
None	NDa	ND	ND	ND	
Hyoscyamine	1.60	0.11	ND	0.06	88.5
Hyos-OH	ND	1.23	ND	0.21	72.0
De-hyos	ND	ND	1.75	0.17	96.0
Scopolamine	ND	ND	ND	1.65	82.5

Approximately 0.2 g fresh weight of the shoot cultures were incubated in light for one week with 2 µmol/10 ml of the tropane alkaloid. Alkaloids in the cells and in the medium were determined separately by gas–liquid chromatography and the amounts were combined. Values are the mean of three flasks. Hyos-OH, 6β-hydroxyhyosyamine; De-hyos, 6,7-dehydrohyosyamine. ND, not detected.

Ornithine to N-methylputrescine

Tropine is synthesized from ornithine via N-methylputrescine (Fig. 2). Earlier tracer experiments established that feeding [2-^{14}C]ornithine yields hyoscyamine labelled only at the C-1 bridgehead carbon of tropine (Leete 1962). This result, together with the fact that δ-N-methylornithine is incorporated to hyoscyamine without cleavage of its N-methyl group (Ahmad & Leete 1979), suggests a biosynthetic pathway which proceeds: ornithine → δ-N-methylornithine → N-methylputrescine. We searched for the N-methyltransferase and the decarboxylase indicated by this proposed pathway in our root cultures, but without success. Instead, we found strong activities of putrescine N-methyltransferase (PMT) only in the tissues that produce alkaloids. Arginine decarboxylase (ADC) activity was about twice as high as ornithine decarboxylase (ODC) activity in H. $albus$ root cultures. ADC, ODC and arginase activities were highest after three to four days in culture, whereas PMT activity was highest at around the sixth day in culture. Thereafter, the free N-methylputrescine content in the root cells increased, followed by an increase in hyoscyamine content. Our feeding studies with [1,4-^{14}C] putrescine, DL-[5-^{14}C]ornithine and L-[1,3-^{3}H]arginine, in combination with administration of an irreversible inhibitor of ODC, such as DL-α-difluoromethylornithine, also indicated that most, if not all, of N-methylputrescine in our $Hyoscyamus$ root cultures is synthesized via symmetrical putrescine (unpublished data). At present we cannot reconcile the previous asymmetrical scheme with our recent results. Further studies on these enzymes may give clues which will clarify this discrepancy.

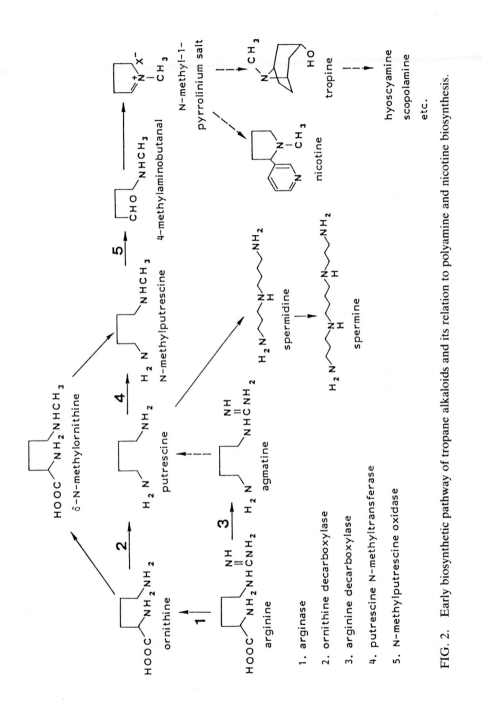

FIG. 2. Early biosynthetic pathway of tropane alkaloids and its relation to polyamine and nicotine biosynthesis.

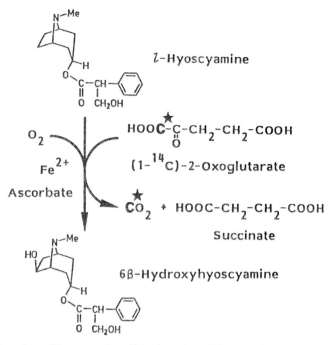

FIG. 3.   Reaction of hyoscyamine 6β-hydroxylase. The reaction shown here is that
when labelled 2-oxoglutarate is used.

Hyoscyamine to scopolamine

Scopolamine is synthesized from hyoscyamine via 6β-hydroxyhyoscyamine
(Fig. 1). 6,7-Dehydrohyoscyamine is also converted to scopolamine when fed
to the *Datura* scions (Fodor et al 1959). The previously proposed pathway of
this epoxidation was either: hyoscyamine → 6,7-dehydrohyoscyamine →
6β-hydroxyhyoscyamine → scopolamine (Waller & Nowacki 1978); or hyoscyamine → 6β-hydroxyhyoscyamine → 6,7-dehydrohyoscyamine → scopolamine (Leete 1979). We found that the first reaction of this epoxidation is
catalysed by a 2-oxoglutarate-dependent dioxygenase, hyoscyamine 6β-
hydroxylase (Hashimoto & Yamada 1986). The hydroxylase incorporates one
atom of molecular oxygen into the 6β-position of *l*-hyoscyamine, while the
other atom of molecular oxygen is simultaneously used to decarboxylate
2-oxoglutarate to succinate (Fig. 3). A 1:1 stoichiometry was shown between
the hyoscyamine-dependent formation of $^{14}CO_2$ from [1-^{14}C]2-oxoglutarate
and the hydroxylation of hyoscyamine. Ferrous ions, ascorbate and catalase
are needed for maximal activity. The hydroxylase was purified 310-fold from
H. niger root cultures that produce scopolamine as a major alkaloid. The
purified enzyme showed the typical characteristics of a 2-oxoglutarate-

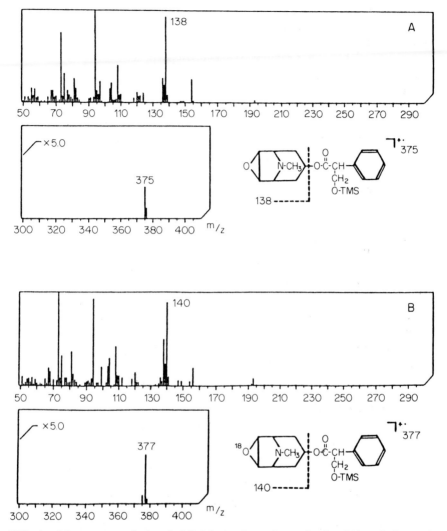

FIG. 4. Mass spectra of trimethylsilyl derivatives of scopolamine (A) and of scopolamine formed *in vivo* from [6-^{18}O]6β-hydroxyhyoscyamine (B).

dependent dioxygenase (EC 1.14.11.-) with regard to its cofactor requirements and its response to most inhibitors, whereas the inhibition studies with pyridine dicarboxylates and *o*-dihydroxyphenyl derivatives indicated that the binding sites for 2-oxoglutarate of the hydroxylase are similar but not identical to those of other 2-oxoglutarate-dependent dioxygenases (Hashimoto & Yamada 1987a). The purified hydroxylase not only hydroxylated various hyoscyamine analogues at the C-6 position of the tropine moiety, but also epoxidized 6,7-dehydrohyoscyamine to scopolamine. There appears to be no difference in the molecular weight and cofactor requirements of the epoxidase and the hydroxylase. Furthermore, the epoxidase activities were found in the same fractions as the hydroxylase activities during purification. We thus conclude that the two reactions are catalysed by a single enzyme, hyoscyamine 6β-hydroxylase.

To determine whether 6,7-dehydrohyoscyamine is, in fact, an *in vivo* precursor of scopolamine, [6-^{18}O]6β-hydroxyhyoscyamine was prepared by the reaction of hyoscyamine 6β-hydroxylase under $^{18}O_2$. The alkaloid labelled with ^{18}O was fed to *Duboisia* shoot cultures, then the scopolamine formed was analysed by gas chromatography — mass spectometry (Hashimoto et al 1987). The scopolamine retained ^{18}O at its epoxide oxygen (Fig. 4), which clearly demonstrates that 6,7-dehydrohyoscyamine is not involved in the scopolamine biosynthesis. This explains why 6,7-dehydrohyoscyamine has never been isolated from plants. The conversion of the unsaturated alkaloid to scopolamine when fed to plant tissues seems to be catalysed by the epoxidase activity in hyoscyamine 6β-hydroxylase.

The *in vitro* conversion of 6β-hydroxyhyoscyamine to scopolamine was demonstrated using the crude enzyme preparations from *H. niger* root cultures (unpublished data). The reaction requires the cofactors for 2-oxoglutarate-dependent dioxygenases and proceeds with retention of the hydroxyl oxygen. At present, the epoxidase activity is relatively weak and our attempts to purify the enzyme have been largely unsuccessful. Still unknown factors may be involved in this reaction.

Other interesting reactions

There are many other interesting, although poorly understood, reactions in the tropane alkaloid biosynthesis (see Fig. 1). A stereospecific condensation of *N*-methyl-1-pyrrolinium cation and acetoacetyl CoA has been postulated to give (2R)-hygrine (Leete 1979). But we recently found that a simple mixture of *N*-methyl-1-pyrrolinium cation and acetoacetic acid (but not its CoA ester) in the absence of tissues readily gives hygrine under physiological conditions (unpublished data). No enzyme extracts from root cultures have so far enhanced this chemical reaction. Chemically synthesized hygrine should be a racemic mixture but only the (2R)-isomer of hygrine would be distinguished by an enzyme at the next reaction step to give ultimately the tropane ring system.

Tropic acid is formed by the intramolecular rearrangement of the side chain of phenylalanine (Leete et al 1975). The actual biosynthetic reactions involved in this interesting rearrangement have not been identified. Esterification of tropine and tropic acid to give hyoscyamine is an important reaction in the pathway but has not been demonstrated conclusively at the enzyme level. Incubation of tropoyl CoA (Gross & Koelen 1980), tropine and enzyme extracts from our cultured roots did not lead to the production of hyoscyamine (unpublished data). These reactions can be studied in detail using enzyme preparations from alkaloid-producing root cultures.

Conclusions

Although tracer and feeding experiments at the level of the whole plant have provided valuable information on the biosynthetic pathway of tropane alkaloids, final confirmation of the pathway must await the discovery and characterization of the enzymes responsible for each biosynthetic step. The danger in relying too much on feeding experiments alone is illustrated by the demonstration that 6,7-dehydrohyoscyamine has been mistakenly regarded as a precursor of scopolamine. Studies of the enzymes involved and, in the future, of their genes will not only establish the biosynthetic pathway but also reveal the mechanisms of its regulation. Productive root cultures will continue to be useful materials for these studies. Unique properties of *Duboisia* shoot cultures should also be valuable for some feeding experiments.

Acknowledgements

We thank T. Endo, Y. Yukimune and J. Kohno for their technical assistance during this work.

References

Ahmad A, Leete E 1970 Biosynthesis of the tropine moiety of hyoscyamine from δ-N-methylornithine. Phytochemistry (Oxf) 9:2345–2347

Evans WC 1979 Tropane alkaloids of the Solanaceae. In: Hawkes JG et al (eds) The biology and taxonomy of the Solanaceae. Academic Press, London, p 241–254

Fodor G, Romeike A, Janzso G, Koczor I 1959 Epoxidation experiment *in vitro* with dehydrohyoscyamine and related compounds. Tetrahedron Lett 7:19–23

Griffin WJ 1979 Organization and metabolism of exogenous hyoscyamine in tissue cultures of a *Duboisia* hybrid. Naturwissenchaften 66:58

Gross GG, Koelen KJ 1980 Chemical synthesis of tropoyl Coenzyme A. Z Naturforsch Sect C Bio Sci 35:363–366

Hashimoto T, Yamada Y 1983 Scopolamine production in suspension cultures and redifferentiated roots of *Hyoscyamus niger*. Planta Med 47:195–199

Hashimoto T, Yamada Y 1986 Hyoscyamine 6β-hydroxylase, a 2-oxoglutarate-dependent dioxygenase, in alkaloid-producing root cultures. Plant Physiol 81:619–625

Hashimoto T, Yamada Y 1987a Purification and characterization of hyoscyamine 6β-hydroxylase from root cultures of *Hyoscyamus niger* L. Eur J Biochem 164:277–285

Hashimoto T, Yamada Y 1987b Effects of culture conditions on tropane alkaloid formation in *Hyoscyamus niger* suspension cultures. Agric Biol Chem 51:2769–2774

Hashimoto T, Yukimune Y, Yamada Y 1986 Tropane alkaloid production in *Hyoscyamus* root cultures. J Plant Physiol 124:61–74

Hashimoto T, Kohno J, Yamada Y 1987 Epoxidation *in vivo* of hyoscyamine to scopolamine does not involve a dehydration step. Plant Physiol 84:144–147

Leete E 1962 The stereospecific incorporation of ornithine into the tropine moiety of hyoscyamine. J Am Chem Soc 84:55–57

Leete E 1979 Biosynthesis and metabolism of the tropane alkaloids. Planta Med 36:97–112

Leete E, Kowanko N, Newmark RA 1975 Use of carbon-13 nuclear magnetic resonance to establish that the biosynthesis of tropic acid involves an intramolecular rearrangement of phenylalanine. J Am Chem Soc 97:6826–6830

Romeike A 1978 Tropane alkaloids-occurrence and systematic importance in angiosperms. Bot Not 131:85–96

Waller GR, Nowacki EK 1978 Alkaloid biology and metabolism in plants. Plenum, New York, p 71

Yamada Y, Hashimoto T 1982 Production of tropane alkaloids in cultured cells of *Hyoscyamus niger*. Plant Cell Rep 1:101–103

DISCUSSION

Zenk: I think Professor Yamada you should be congratulated on elucidating these pathways. This has been a long-standing problem in plant biochemistry and I think you have just perfectly solved it. We have been ourselves engaged in these experiments but we have been overrun by your group at least by one year. What is really interesting is that until now one has failed to get suspension cultures to produce scopolamine or hyoscyamine in good quantities, so that one had to turn to root cultures. I would like to point out that there is a long-forgotten patent, which in Germany was the first patent in plant biotechnology, filed by Dr H. Metz from the Merck company (Metz 1964) on the procedure for the cultivation of differentiated root tissue. Don't worry it's long expired, but it's interesting that some of the things that are fashionable today reflect comparatively old technology. The important question which now remains is the esterification of tropine with L-tropic acid to form hyoscyamine. Did you ever consider non-aqueous ester transfer, either by a plant enzyme or by microbial enzymes, to do this esterification? Tropine is easy to synthesize, tropic acid is available, so the only thing you need is this ester-forming enzyme to have the whole pathway potentially commercially available.

Yamada: We synthesized tropoyl Coenzyme A, in collaboration with Professor H. Inoue, Faculty of Pharmaceutical Science, Kyoto University, in an attempt to use its Co A energy for esterification with tropine. Unfortunately, we have not yet found a preparation containing the esterification enzyme, even though extracts of several plant species have been tested (T. Hashimoto, H. Inoue, B. Dräger, Y. Yamada, unpublished work). We have not tried non-aqueous reaction conditions. As for bacterial esterases, such as tropine esterase (EC 3.1.1.10), I think that they only break ester bonds and do not form them. I agree that this is one of the most important reactions in the tropane alkaloid biosynthesis.

Barz: With regard to the esterification of tropic acid, you are only looking for the Coenzyme A ester. Does it necessarily have to be a Co A ester? Have you considered a glucose ester? There are several examples in the literature where a glucose ester is the energy-rich compound involved in the biosynthesis of an ester.

Yamada: Yes, we also considered that tropoyl glucose may work for ester-ification with tropine, so UDP-glucose, tropic acid and tropine were incubated with *Hyoscyamus* enzymes. This did not work either (T. Hashimoto, Y. Yama-da, unpublished work). At this moment, we don't know what kind of energy we can supply to drive this reaction. Do you have any suggestions?

Barz: The tropic acid glucose ester can be synthesized chemically; you could try that in your system.

Fowler: With reference to Professor Zenk's point on the non-aqueous en-vironment, is there any evidence of a membrane bound system, and if so, have you tried to recover it as a pellet by centrifugation.

Yamada: You mean a lipophilic membrane?

Fowler: Yes, that is a possibility. You mentioned two possible reactions of the enzyme system. Have you analysed the purified protein by SDS-PAGE? Is it a single protein band?

Yamada: All the data reported here were obtained with highly purified, but not homogeneous, hyoscyamine 6β-hydroxylase. However, we recently suc-ceeded in purifying the enzyme to homogeneity, as judged by SDS-PAGE (S. Okabe, T. Hashimoto, Y. Yamada, unpublished work).

Fowler: I was intrigued by the enzyme/protein patterns, which indicated slight differences in substrate specificity.

Rhodes: I would like to make a comment about Professor Yamada's experi-ence with transformed root cultures. Our experience with *Hyoscyamus niger* is limited to the one or two cultures that we have developed, but we found, as you did, that the production rate was quite low. However, in related species such as *Datura stramonium*, we have cultures which are producing up to 2% of their dry matter as hyoscyamine. We have a culture derived from a *Datura* hybrid which is producing almost as much scopolamine, so I think that in solanaceous species it is possible to have transformed root cultures with relatively high rates of alkaloid production.

Yamada: We have screened many solanaceous species for the best source of the enzyme hyoscyamine 6β-hydroxylase, but not for productivity of total alkaloids. Root cultures of *H.niger*, although relatively low in alkaloid content, grow well and have the highest enzyme activity per fresh root mass (Hashimoto & Yamada 1986). Probably the high ratio of scopolamine to hyoscyamine in these root cultures is a reflection of their high hydroxylase activity. Other root cultures, such as *Datura* and *Atropa*, may have a higher total alkaloid content, but their scopolamine:hyoscyamine ratios are low, and thus their hydroxylase activities are also low. If you are looking at the production of total tropane alkaloids, root cultures of other species may be preferable. Indeed, *Duboisia leichhardtii* is known to have a high alkaloid content as well as a relatively high scopolamine:hyoscyamine ratio. Sumitomo Chemical Company in Takarazuka, Japan, with which I am affiliated, reported at a meeting of the Japanese Association of Plant Tissue Culture (1987) that they obtained transformed roots of *D. leichhardtii* that produced about 2% of their dry weight as tropane alkaloids.

Rhodes: Does the *Duboisia* species you use produce nicotine as well as the tropanes and if so is competition between the two groups of compounds for the common precursor, *N*-methyl pyrrolinium, ultimately a limitation to the maximum production of tropanes?

Yamada: All *Duboisia* species produce nicotine.

Rhodes: Yes, but in the particular one that you used, did you have both classes of compounds being produced? Was there potentially competition for *N*-methyl pyrrolinium in the two pathways.

Yamada: Tobacco alkaloids have not been analysed in the transformed *Duboisia* roots, but our untransformed *Duboisia* roots produce both nicotine and tropane alkaloids (Endo & Yamada 1985). We don't know if there is competition for common precursors between the nicotine and tropane alkaloid biosynthetic pathways in *Duboisia*.

Galun: *H. sinensis* is considered to be very high in alkaloids (it's actually a toxic plant). You could try that.

Yamada: Does it grow in Japan?

Galun: I will tell you the source. My other remark is that for scopolamine, the highest quantity that we achieved in leaves of *Datura* was approximately 3% of dry weight. This was obtained in *D. sanguinea*, which is also called the *Bruginensia*. The leaves contain very high amounts of scopolamine after selection and seeds can be obtained free from us, if anyone likes to have them.

Zenk: We heard a little about the content of scopolamine in the roots, values of 0.1%, 0.2%. What is the yield in mg/l of culture? I think this is the fairest way to express yield. This question is also directed to Dr Rhodes.

Rhodes: With *D. stramonium* we have produced up to 0.5 g/l of hyoscyamine in fermenters in which the transformed roots were grown to high densities (>25 g dry weight/l).

Zenk: This is a growth rate of about four weeks?

Fowler: That is a key question, the timing of the achievement of that yield is crucially important.

Yamada: Our root culture system has not been optimized for high productivity: for example, high-density cultures have not been tested. But I can give you some preliminary figures: *H. niger* roots, 1.5mg scopolamine/75 ml of culture medium/2 weeks of incubation; *H. albus* roots, 1.2mg hyoscyamine/25 ml/

week; transformed *D.leichhardtii* roots (Mano, Ookawa, Yamada, unpublished work), 3.9mg scopolamine/50 ml/4 weeks.

We are using a two stage culture method of *H.niger* in a culture medium with auxin. Auxin promotes the root growth but inhibits scopolamine synthesis. Then we transfer the roots to a medium lacking auxin and culture them for one week (Hashimoto et al 1986). This medium enhances the enzyme activities involved in scopolamine biosynthesis.

Tabata: Do you find high activity of hyoscyamine 6β-hydroxylase only in the roots, not in the shoots or culture cells? We have detected scopolamine in the callus cultures of *Datura innoxia* only after roots had been initiated from these cultures.

Yamada: High levels of hydroxylase activity are present only in the roots, but shoots of *Duboisia myoporoides* do contain a relatively weak hydroxylase activity (Hashimoto & Yamada 1986). The reason the *Duboisia* shoot cultures contain no alkaloid is that they do not synthesize hyoscyamine, a precursor of scopolamine.

Tabata: Can you tell us about the substrate specificity of this particular enzyme? Does it hydroxylate other related compounds, such as tropine, nor-hyoscyamine or apo-hyoscyamine?

Yamada: We have worked with about 30 analogues of hyoscyamine. We can compare the ability of these alkaloids to act as substrates by taking the hydroxylation rate of *l*-hyoscyamine as 100%: tropine is not hydroxylated; *l*-norhyoscyamine = 81%; and apoatropine = 45% (Hashimoto & Yamada 1987). Although the enzyme requires the basic structure of the tropane alkaloid ester for the reaction to occur, it has relatively low specificity for alkaloid substrates.

Zenk: It's intriguing that the synthetic compound 6,7-dehydrohyoscyamine is hydroxylated or epoxidized by your enzyme so efficiently. In our root culture systems, if we feed it as a chemical substance it is the substance which is best transformed to scopalamine. One can get very high conversion of dehydrohyoscyamine to scopalamine, better than hyoscyamine or 6-hydroxy hyoscyamine.

Yamada: We have obtained similar results (Hashimoto et al 1987). This is probably because both hyoscyamine 6β-hydroxylase and an epoxidase are needed to convert hyoscyamine to scopolamine, but the hydroxylase alone can convert 6,7-dehydrohyoscyamine to scopolamine. 6,7-Dehydrohyoscyamine is as good a substrate as hyoscyamine for the hydroxylase *in vitro* (Hashimoto & Yamada 1987).

Barz: You showed the condensation step to nicotine, do you have any information on this enzyme reaction?

Yamada: At this moment we don't have any information.

Rhodes: The possible alternative route from ornithine to putrescine via *N*-methyl ornithine was shown on one of your slides in brackets. In many text books this is shown as the major route. Do you have evidence that this route is of less importance in the tissues you have studied?

Yamada: Once this kind of scheme has been postulated then I think everyone thinks that it has been proven. *Hyoscyamus* species might be exceptional but so far from all our studies (Hashimoto, Yukimure, Yamada, unpublished work), we are convinced that a pathway *via* symmetrical diamine putrescine operates *in vivo*.

References

Endo T, Yamada Y 1985 Alkaloid production in cultured roots of three species of *Duboisia*. Phytochemistry 24:1233–1236

Hashimoto T, Yamada Y 1986 Hyoscyamine 6β-hydroxylase, a 2-oxoglutarate-dependent dioxygenase, in alkaloid-producing root cultures. Plant Physiol 81:619–625

Hashimoto T, Yamada Y 1987 Purification and characterization of hyoscyamine 6β-hydroxylase from root cultures of *Hyoscyamus niger* L. Eur J Biochem 164:277–285

Hashimoto T, Kohno J, Yamada Y 1987 Epoxidation *in vivo* of hyoscyamine to scopolamine does not involve a dehydration step. Plant Physiol 84:144–147

Hashimoto T, Yukimune Y, Yamada Y 1986 Tropane alkaloid production in *Hyoscyamus* root cultures. J Plant Physiol 124:61–75

Metz H 1964 Submerged culture of plant roots. 2nd International Fermentation Symposium, Chem & Ind Abstr p552

Biotechnological approaches to the production of isoquinoline alkaloids

M.H. Zenk, M. Rüffer, T.M. Kutchan and E. Galneder

Lehrstuhl für pharmazeutische Biologie, Universität München, Karlstraße 29, D-8000 München 2, Federal Republic of Germany

Abstract. Isoquinoline alkaloids are a major group of medicinally important compounds. Their structural complexity and diversity have made them an interesting subject of study for chemists and biochemists alike. Although these natural products have significant pharmaceutical value, quantities sufficient to meet commercial demand are in some cases difficult to obtain due to limited or expensive supplies of the plant material from which they are derived. Many of these products have sophisticated structures, including chiral centres, which make their chemical synthesis difficult and impractical. Alternative methods of production are therefore essential. Production of natural compounds from plant cell cultures has been successful. Other means of achieving specific products are the biotransformation of drug precursors by immobilized plant cells or biomimetic enzyme-assisted synthesis. For more general application, we are now interested in using the techniques of molecular genetics to develop methods for the microbial production of the enzymes of plant secondary metabolism.

1988 Applications of plant cell and tissue culture. Wiley, Chichester (Ciba Foundation Symposium 137) p 213–227

Isoquinoline alkaloids constitute a major group of plant-derived pharmaceuticals. As compiled by Farnsworth (1984), there are a total of 46 plant-derived drugs widely employed in Western medicine, of which six are isoquinolines. In the United States of America, the consumer paid about $8 billion during 1980 for prescription drugs of which the active principles were obtained from plants. Furthermore, there are 34 plant-derived drugs used widely but not generally in Western medicine, out of which six are isoquinolines (Farnsworth 1984). Therefore, of all the 80 plant-derived drugs used in the world, 16% are isoquinoline alkaloids. Together with the indole alkaloids, they are the most important class of therapeutic agents obtained from plants. Their medicinal value, chemical complexity and structural diversity make them an ideal group of natural products for the study of biosynthesis and biotechnological production. The world population has reached five billion this year (1987), by the turn of the millenium it is expected to double again. This means that there will be an increasing demand for any natural product, but particularly for the indispensible and renewable therapeutic plant-derived drugs. In order to secure a stable supply of these agents in the future and to make industrial nations independent of political and economic crises affecting the drug supply, biotechnological processes for the production of these important compounds have to be explored.

The potential to achieve this goal rests presently with cell and organ culture of isoquinoline-producing plants, as well as with enzyme technology to catalyse those steps in the synthesis of natural products which are neither chemically nor economically feasible. Plant cell cultures are also an interesting tool for the discovery of new compounds with pharmacological or other biological activities, either for use as such or as lead structures for further chemical improvement. In this paper we discuss the current state of development of these techniques and possibilities for their application in the future.

Plant cell cultures

In a systematic study (Zenk et al 1975) we successfully demonstrated for the first time that completely dedifferentiated cell suspension cultures of a higher plant (*Morinda citrifolia*) can produce secondary metabolites (anthraquinones) in the range of a gram per litre medium during a cultivation time of 2–3 weeks. The optimized cell culture surpassed the field-grown intact plant in its metabolite content by a factor of about 10. This opened the way for the exploration of industrial application of plant cell culture (Zenk 1978).

The strategies involved the following steps: 1) development of an analytical or pharmacological test system which allows the screening of thousands of plants and cell culture clones (such as an immunosassay or receptor assay system, visible or UV-examination for the desired metabolite). 2) Selection of individual plants containing a higher than average concentration of the desired chemical. 3) Establishment of callus and subsequently cell suspension cultures in a growth medium of the high-producing plant. 4) Development of a production medium which allows, by variation of the quantity and quality of macro- and micronutrients, hormones or precursors, the maximal production of the desired metabolite. 5) Cloning of single cells, small cell aggregates or protoplasts to obtain stable, high-producing cell lines. 6) Scale up of the cultures to produce the compounds of interest in the greatest possible yields under constant monitoring for their biochemical and growth characteristics (Zenk et al 1977). The advances which have been made using this and similar strategies for the improvement of plant cell suspension cultures have been excellently and critically reviewed (Fontanel & Tabata 1987).

Spontaneous production. We now turn specifically to the production of isoquinoline alkaloids by plant cell cultures. The occurrence of isoquinoline alkaloids in plant tissue cultures up to 1984 has been summarized by Rüffer

(1985). As a result, we can conclude that all major groups of isoquinoline alkaloids (aporphines, benzophenanthridines, bisbenzylisoquinolines, phthalideisoquinolines, proaporphines, protoberberines, protopines and tetrahydrobenzylisoquinolines) are spontaneously produced in cell cultures of various plants. In general, however, we can assume that in the isoquinoline series all structural types of compounds can be produced, even if some of these compounds are synthesized in only very small amounts. The morphinandienones, having the greatest economic impact, are produced in notoriously low concentrations by cell cultures of *Papaver somniferum* and in most cases their synthetic capacity is lost after a few passages.

An example for the spontaneous production of considerable amounts of protoberberines is found in the genus *Berberis*. Thirty-six species in culture were investigated by us and almost all of these strains were producing jatrorrhizine as the major component, in addition to small amounts of berberine, columbamine and palmatine.

The highlight in the field is, however, the successful economic fermentation of berberine using *Coptis japonica* cells by the Mitsui Petroleum company in Japan (Fujita & Tabata 1987). Using a strain of *C. japonica* developed by Sato & Yamada (1984), and applying the most modern methods of optimization of cell cultures, the Mitsui company obtained yields of seven grams of berberine per litre of medium, which is the world record for any compound ever synthesized by cell suspension culture. The way to achieve this extraordinary yield involved selection of highest yielding protoplasts by a fluorescence-activated cell sorter and subsequent regeneration of these super-producing protoplasts to wall-containing strains, rigorous medium optimization and creation of heterokaryons by protoplast fusion of fast-growing with high-producing clones. This most remarkable achievement stands next to the development of the shikonin process by the same company as a landmark for the biotechnological application of plant cell culture. These achievements have been obtained simply on an empirical basis without any knowledge of the biosynthesis of these alkaloids and their regulation. It should, however, be pointed out that the price of berberine on the world market is only around US $100/kg, a price at which this compound could definitely not be produced by fermentation in Western Europe. Future attempts in the field will have to concentrate on the development of alkaloid production medium to increase some of the more interesting isoquinoline alkaloid productions to the gram per litre range. The two-stage fermentation system (growth medium followed by production medium) introduced by us (Zenk et al 1977) for plant cell cultures was successfully employed by Fujita et al (1981) and can also be applied to isoquinoline-containing cell cultures. In each case the production medium must, unfortunately, be optimized for the specific plant system used.

Screening for new compounds. It has been known for many years that dedifferentiated plant cells produce compounds which are not found in the differentiated plant. It is therefore expected that cell cultures will contain entirely new compounds with biological activity, as well as those already known. The possibility of patentable new products and processes may overcome some of the problems which defer industry from exploring plant-derived products (Farnsworth 1984).

An innovative new development was introduced by Arens et al (1982), in which large numbers of plant cell cultures were screened using highly sensitive receptor binding assays, e.g. opiate receptor from rat brain. This screening programme identified completely new pharmacologically active compounds (Arens et al 1982). A similar screening system using a benzodiazepine receptor assay (from rat brain) with cell extracts of *Aristolochia grandifolia* led to the isolation of a highly active compound (affinity to receptor 5×10^{-7} M), which was identified as the isoquinoline derivative, 2-hydroxy-1-methoxy-dibenzoquinoline-4,5-dione (W. Wehrli et al, Research department CIBA-GEIGY Ltd Basel, Switzerland, unpublished). It was not previously known that this type of compound can react with the benzodiazepine receptor and this information may serve as a lead for future synthesis.

Immobilization of cells. Immobilization of plant cells was successfully employed for the first time by Brodelius et al in 1979. This method can, however, only be applied to cell systems which catalyse biotransformations and excrete the product into the medium. A cell line of *Thalictrum minus* var. *hypoleucum* that excretes berberine into the medium has been described by Kobayashi et al (1987). Cells entrapped in calcium alginate released most of the berberine synthesized into the liquid medium and a newly devised bioreactor was used for the purpose of producing berberine. It will be most interesting to learn what physiological changes have taken place in that particular *Thalictrum* species, since all of the 45 species of *Thalictrum* examined by us do not show this phenomenon but store the berberine formed in their vacuoles.

Enzymology of isoquinoline alkaloid formation

The enzymes. From a biotechnological view point there is a threefold interest in the enzymes catalysing the formation of secondary metabolites: elucidation of the true precursors for a given end-product, discovery of the rate-limiting steps (if there are any) in a biosynthetic pathway, and use of the isolated enzymes for biomimetic syntheses. Again, cell cultures have proven to be a more reliable and richer source of enzymes than the differentiated plants. Table 1 shows a comparison of the enzyme activities involved in berberine synthesis found in a 2.5-year-old *Berberis* plant as compared to a nine day cell suspension culture of the same plant. In all but one case, the

TABLE 1 Comparison of the enzymes involved in berberine biosynthesis in a 2.5 year old plant root and a nine day old cell suspension culture of Berberis

Enzyme	Optimum pH	Particle bound	Stereo-specificity	Plant root 2.5 years old (pkat/g DW)	Cell suspension culture 9 days old (pkat/g DW)	Increase (cell culture /root)
1 (S)-Norcoclaurine-synthase	7.8	No	Yes	380	70	0.2
2 Norcoclaurine-6-O-methyltransferase	7.5	No	No	90	189	2
3 Coclaurine-N-methyltransferase	7.7	No	No	10	1200	120
4 N-Methylcoclaurine-3'-hydroxylase[a]	—	—	—	—	—	—
5 (S)-3'-Hydroxy-N-methylcoclaurine-4'-O-methyltransferase	7.5	No	Yes	11	309	28
6 Berberine bridge enzyme	8.9	Yes	Yes	57	65	1.1
7 (S)-Scoulerine-9-O-methyltransferase	8.9	Yes	Yes	41	204	5
8 (S)-Tetrahydroprotoberberine oxidase	8.9	Yes	Yes	350	2400	7
9 Berberine synthase	8.9	Yes	—	10	123	3

[a] Discovered but not yet characterized.

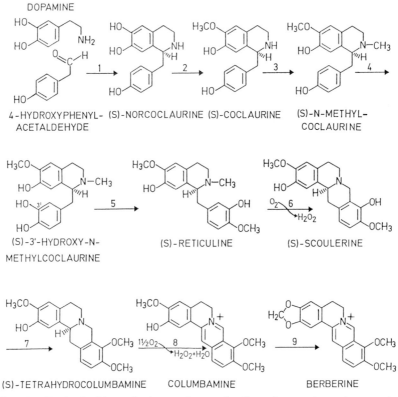

FIG. 1. Revised biosynthetic pathway leading from dopamine and 4-hydroxyphenylacetaldehyde to berberine in *Berberis*.

enzyme concentration was much greater in the cell culture than in the differentiated plant. Studies in cell cultures of the *Berberidaceae* and *Papaveraceae* families have revealed all the enzymes involved in the biosynthesis of protoberberine. The pathway, as we consider it now, is depicted in Fig. 1. If we analyse the specific activities of the enzymes involved in this pathway, we cannot identify the rate-limiting catalyst. The product yield by *Berberis*, *Eschscholtzia* or *Thalictrum* strains could not be increased by supplying precursors such as coclaurine or reticuline. Use of colourless variant strains was also unsuccessful, in these cases *all* the enzyme activities were lower than in the producing variants. The exploration of biosynthetic pathways at the enzyme and metabolite level in these non-producing variant strains is important, since if we can learn from these examples how non-producing strains are deregulated, we may be able to form hypotheses about how to overcome this problem.

In our opinion there are two enzymes in this pathway which could be of interest for biotechnology. One is the berberine synthase (enzyme 9, Table

1). It catalyses the formation of the methylenedioxy bridge, starting with columbamine as substrate. This is a reaction which cannot be conducted by organic chemistry. The second enzyme is the (S)-tetrahydroprotoberberine oxidase (enzyme 8, Table 1), which, in the presence of 1.5 moles of oxygen, stereospecifically catalyses the aromatization of ring C in this alkaloid, simultaneously producing, per mole of substrate, one mole of hydrogen peroxide and one mole of water. This reaction in itself is trivial. The aromatization can be conducted chemically quantitatively and at low cost. We have, however, discovered that this oxidase, a flavoprotein, catalyses the stereospecific dehydrogenation of (S)-benzylisoquinolines to their corresponding 1,2-didehydro derivatives, which in turn can be reduced to the enantiomeric mixture. By recycling we succeeded in the transformation of a racemic mixture of reticuline into the pure compound with the (R)-configuration which is an important precursor for the NIH-codeine synthesis (Amann et al 1987). The half-life of this enzyme was increased from four days for the soluble enzyme to 200 days at room temperature when immobilized, making it a valuable analytical and biotechnological tool.

Since *Coptis japonica* has been such a successful example of berberine production at the commercial level, we also checked the course of biosynthesis of the alkaloid in this plant. Only one enzyme from this pathway and this species has previously been reported (Yamada & Okada 1985), a crude oxidase converting (S)-tetrahydroberberine into berberine. According to these authors, this reaction does not produce hydrogen peroxide, the four hydrogen atoms of the substrate are removed by the production of two water molecules. We (Galneder et al 1988) have now systematically investigated the biosynthesis of berberine in *Coptis*, compared it to *Berberis* and, much to our surprise, have found that the tetrahydroprotoberberine oxidase of *Coptis* is indeed different from that in *Berberis*. The major difference is the substrate specificity. The *Coptis* oxidase does not dehydrogenate benzylisoquinolines; it is not a flavoprotein but contains a metal (iron). Furthermore, in *Coptis*, the methylenedioxy group is formed at an earlier stage. In *Berberis* this step is catalysed by the last enzyme of the pathway. Therefore the *Coptis* pathway to berberine has to be formulated as follows: (S)-reticuline → (S)-scoulerine → (S)-tetrahydrocolumbamine → (S)-tetrahydroberberine→berberine. The previously established pathway to berberine in *Berberis* differs considerably in the terminal steps: (S)-reticuline → (S)-scoulerine → (S)-tetrahydrocolumbamine → columbamine → berberine. The *Coptis* pathway is also used in the four *Thalictrum* species that we have investigated so far. It is of considerable interest that Nature has developed for one and the same compound, berberine, pathways which differ completely in two steps. Consideration of the close taxonomical relation between the family *Ranunculaceae*, to which *Coptis* and *Thalictrum* belong, and the family *Berberidaceae* (both families belong to the order *Ranunculales*), makes this result even more

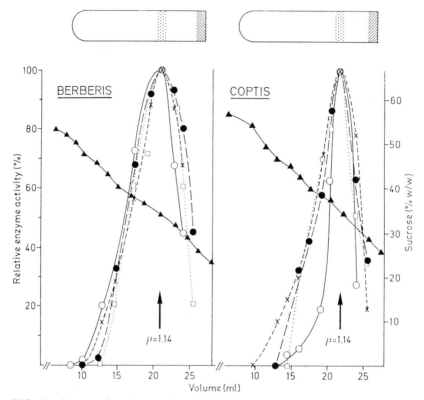

FIG. 2.   Sucrose density centrifugation on a linear gradient (20–60% w/w) of crude extracts from cell cultures. *Berberis*: berberine bridge enzyme (●--●); (S)-scoulerine-9-O-methyltransferase (□····□); (S)-tetrahydroprotoberberine oxidase (X--X); berberine synthase (O—O); sucrose concentration (▲—▲). *Coptis*: berberine bridge enzyme (●--●); (S)-scoulerine-9-O-methyltransferase (□····□); (S)-canadine synthase (O—O); (S)-canadine oxidase (X--X); sucrose concentration (▲—▲).

surprising. This fact warns us not to generalize pathways for secondary metabolites, even if the metabolic route has been completely established at the enzyme level for one species.

Enzyme compartmentalization.   A routine check on the supposed cytoplasmic localization of the nine enzymes involved in berberine biosynthesis in *Berberis* led to the discovery of a new vesicle in which four of the enzymes (Table 1, enzymes 6-9) are exclusively located. None of the enzymes involved in the formation of the protoberberine nucleus are found in the cytosol. It is not surprising that the pathway from (S)-reticuline to berberine in *Coptis* (as well as *Thalictrum*) is also exclusively localized in the same kind of vesicle (Fig. 2). This points to the general principle that the quaternary alkaloids of the protoberberine series are not able to pass through biological membranes.

In order to reach their final destination, the whole biosynthetic compartment containing both products and enzymes moves to the tonoplast membrane where the vesicle membrane fuses with the tonoplast membrane, releasing the contents of the vesicle, the quaternary alkaloids, into the vacuole where they are stored. As discussed above, the berberine-excreting strain of *T. minus* (Kobayashi et al 1987) is an interesting exception in that the vesicles, in which undoubtedly, in this case, berberine is synthesized, instead of moving to the tonoplast membrane are directed to the plasmalemma, resulting in this abnormal excretion phenomenon.

Alkaloid storage

In contrast to the mechanism of transport for quaternary protoberberines depicted above, the accumulation of tertiary alkaloids is achieved by a completely different mechanism. Using isolated vacuoles derived from suspension cells of *Fumaria capreolata*, it could be shown that tertiary alkaloids are taken up, not by an ion trap mechanism but via a carrier-mediated, ATP-dependent system (Deus-Neumann & Zenk 1986). This uptake system is absolutely specific for alkaloids indigenous to the plant from which the vacuoles were isolated. On the other hand, alkaloid sequestration in *Papaver somniferum* latex seems to follow a different mechanism, being also highly specific but not energy dependent and with no carrier system being involved (Homeyer & Roberts 1984). There seems to be a fundamental difference in the storage mechanism of the normal vacuole and the specialized latex vesicles. It would be of great academic but also practical interest to identify and re-introduce the vacuolar carrier system into tonoplast membranes, in an attempt to generate plant systems with a higher than normal capacity to store alkaloids and to explore the regulatory role by which a vacuole, having reached its genetically fixed capacity of alkaloid storage, signals the cytoplasm to stop further synthesis.

Genes for alkaloid synthesis

The expression of a plant gene coding for a secondary product, the peptide thaumatin, in a bacterium (see Eden & van der Wel 1985) opened a new area for plant biotechnology. It will be of great importance to explore the possibility of expression, amplification and large-scale isolation of those plant-derived enzymes which catalyse reactions inaccessible to organic chemists. Enzyme biotechnology is at a stage where industrial development is feasible. There are several steps in isoquinoline biosynthesis where the availability of a specific enzyme could, in combination with biomimetic organic synthesis, lead to a commercially interesting product. On the other hand, some of these enzymes that are able to catalyse reactions difficult to achieve by chemical means may require covalently bound cofactors such as flavine, pyrroloquinoline quinone, or metal ions. This could make expression of the biologically active plant protein in a microorganism difficult to achieve. A multitude of chiral plant products and chiral chemical or biochemical precursor molecules could be synthesized by cloning the corresponding plant gene. We envisage the chance to induce mRNA species to facilitate cloning of alkaloid biosynthetic enzymes in a manner similar to that worked out by Hahlbrock and coworkers (Douglas et al 1987) for flavonoid biosynthesis. It will also be necessary to learn about gene organization and gene regulation in pathways of secondary metabolism. One of the future aims in plant biotechnology has to be to express portions of biosynthetic pathways leading to secondary compounds in microorganisms so that the plant product is synthesized by microbes which have much better fermentation characteristics than plants. In order to achieve this goal, we have, at present, to concentrate on the investigation of biosynthetic pathways leading to useful products. We have to identify, characterize and sequence the gene products, those catalytically active proteins involved in the formation of some of Nature's most fascinating molecules, before we can turn to gene technology.

Final remarks

That plant biotechnology will develop is certain, with the development of an estimated dozen commercial plant cell fermentation systems by the end of the century. Enzyme and gene technologies will soon become available, which will have a major impact on the production of agriculturally and medicinally useful plants. We should, however, always remember that the final criterion governing the introduction of all these new techniques into industry is the economic aspect. Only if the plant product under consideration can be produced commercially at a price equal to or lower than that of the field-grown goods will the biotechnological development be given a chance.

Acknowledgements

This work was supported by the Deutsche Forschungsgemeinschaft Bonn (SFB 145) and the Fonds der Chemischen Industrie.

References

Amann M, Nagakura N, Zenk MH 1987 Preparation of dehydrobenzylisoquinolines by immobilized (S)-tetrahydroprotoberberine oxidase from plant cell cultures. Phytochemistry (Oxf) 26:3235–3241
Arens H, Borbe HO, Ulbrich B, Stöckigt J 1982 Detection of pericine, a new CNS-active indole alkaloid from Picralimia nitida cell suspension culture by opiate receptor binding studies. Planta Med 46:210–214

Brodelius P, Deus B, Moosbach K, Zenk MH 1979 Immobilized plant cells for the production and transformation of natural products. FEBS (Fed Eur Biochem Soc) Lett 103:93–97

Deus-Neumann B, Zenk MH 1986 Accumulation of alkaloids in plant vacuoles does not involve an ion-trap mechanism. Planta (Berl) 167:44–53

Douglas C, Hoffmann H, Schulz W, Hahlbrock K 1987 Structure and elicitor or UV-light stimulated expression of two 4-coumarate:CoA ligase genes in parsley. EMBO (Eur Mol Biol Organ) J 6:1189–1195

Edens L, van der Wel H 1985 Microbial synthesis of the sweet tasting plant protein thaumatin. Trends Biotechnol 3:61–64 and literature cited therein

Farnsworth NR 1984 The role of medicinal plants in drug development. In: Krogsgaard-Larsen P et al (eds) Natural products and drug development. Munksgaard, Copenhagen, p 1–28

Fontanel A, Tabata M 1987 Production of secondary metabolites by plant tissue and cell cultures. Present aspects and prospects. In: Boella C et al (eds) Nestlé Research News, p 93–103

Fujita Y, Hara Y, Suga C, Morimoto T 1981 A new medium for the production of shikonin derivatives. Plant Cell Rep 1:61–63

Fujita Y, Tabata M 1987 Secondary metabolites from plant cells — pharmaceutical applications and progress in commercial production, In: Green CE et al (eds) Plant tissue and cell culture. Alan R Liss, New York, p 169–186

Galneder E, Rüffer M, Wanner G, Tabata M, Zenk MH 1988 Alternative final steps in berberine biosynthesis in Coptis and Thalictrum cell cultures. Plant Cell Rep 7: 1–4

Homeyer BC, Roberts MF 1984 Alkaloid sequestration by Papaver somniferum latex. Z Naturforsch Sect C Biosci 39:876–881

Kobayashi Y, Fukui H, Tabata M 1987 An immobilized cell culture system for berberine production by Thalictrum minus cells. Plant Cell Rep 6:185–186

Rüffer M 1985 The production of isoquinoline alkaloids by plant cell cultures. In: Phillipson JD et al (eds) The chemistry and biology of isoquinoline alkaloids. Springer Verlag, Berlin, p 265–280

Sato E, Yamada Y 1984 High berberine-producing cultures of Coptis japonica cells. Phytochemistry (Oxf) 23:281–285

Yamada Y, Okada N 1985 Biotransformation of tetrahydroberberine to berberine by enzymes prepared from cultured Coptis japonica. Phytochemistry (Oxf) 24:63–65

Zenk MH, el-Shagi H, Schulte U 1975 Anthraquinone production by cell suspension cultures of Morinda citrifolia Planta Med Suppl 79–101

Zenk MH, el-Shagi H, Arens H, Stöckigt J, Weiler EW, Deus B 1977 Formation of the indole alkaloids serpentine and ajmalicine in cell suspension cultures of Catharanthus roseus. In: Barz W et al (eds) Plant cell cultures and their biotechnological application. Springer Verlag, Berlin, p 27–43

Zenk MH 1978 The impact of plant cell culture on industry. In: Thorpe TA (ed) Frontiers of plant tissue culture. University of Calgary Press, Calgary, Alberta, Canada, p 1–13

DISCUSSION

Yamada: I would like to say one word personally to Professor Zenk. I fully appreciate your effort to resolve the differences in experimental results between your group and our group. We have been competing with each other, but I think that you have now given the final answer to this problem. I was very impressed by your elucidation of the total metabolic pathway of berberine production.

Fowler: I think that Professor Zenk has put his finger on the key point for application in the future. I am sure that a major way ahead is to pick out discrete enzymes and use these as biotransformation systems within a total process rather than to expect cell culture to achieve everything. In this context, the papers we have heard this morning exemplify the need to improve our understanding of the biochemical pathways, the nature of the enzymes, their location within the cells and so on. Professor Zenk identified an important future direction for plant cell culture as the use of enzymes as discrete catalytic entities outside a cell, rather than perhaps in complete pathways contained inside cells.

Barz: Professor Zenk, you mentioned the strategy of screening callus or suspension cultures of tropical or rare plants by using pharmacological assays to detect interesting compounds. Would you say that elicitation could play a role in this strategy, in that we not only look for compounds which are produced by somaclonal variation (or whatever the heterogeneity is called), but we also try to trigger dormant genes by an external signal?

Zenk: I certainly agree that one can do this. We have tried this sort of approach. Some elicitors are very specific for certain compounds with certain structures, so I think the elicitation process will give an answer for a certain group of chemicals, but I don't think that we will find a super-elicitor that turns on all the secondary genes and thereby increases the activity of a complete pathway or several separate pathways.

Barz: You said that elicitation is not of interest for yield improvement of commercial strains, could you comment on this remark?

Zenk: I wanted to criticise your statement that this property might be exploited for industrial production of useful plant compounds! We have done considerable work on high-yielding strains, which we define as strains that produce over 1 g/l of compound. If you try to elicit them with any of the elicitation systems which are readily available, such as *Phythopthora megaspermia*, yeast or skeroglucan elicitors, we have never found that a high-producing strain can be further increased in yield by elicitation, and so I disagree with your statement.

Cocking: One of the interesting cell biological features of this system is the accumulation of compounds, such as berberine, in vacuoles. Professor Zenk referred to the uptake mechanism specific for the vacuolar membrane. What is the situation regarding the plasma membrane, for example, in relation to uptake into protoplasts? Since we now have the ability to reverse the membrane disposition of protoplast systems, is there an opportunity of getting a better understanding of the capacity of the system to secrete the product, rather than accumulate it in vacuole?

Zenk: That is a very interesting question. We have done only a limited amount of work on this. Dr M Wink in Munich has done some (Mende & Wink 1987) which suggests that there is a fundamental difference between the plasmalemma and the tonoplast. The tonoplast, as I have tried to show, is extremely fussy about what it will take up, while the plasmalemma takes up just about everything. This is good, because if this were not the case, the herbicide industry would be out of business! I think the plasmalemma cannot distinguish chemicals for uptake. Foreign compounds, whether these are herbicides, insecticides or natural compounds (not all of them but some of them) penetrate the plasmalemma relatively easily but then do not proceed further into the vacuole. Only after, for example, hydroxylation of a xenobiotic molecule or glucosylation can the molecule enter the vacuole. So there is an interesting system where the plant cell has to metabolize the compound into a form where it can be stored in the vacuole away from general metabolic enzymes.

Cocking: There is now the capability of isolating vacuoles and fusing the isolated vacuoles with isolated protoplasts, thereby creating modified plasma membranes in which to investigate further these differential control mechanisms between plasma membrane and vacuolar membrane.

Hall: Professor Cocking's point focuses this question of targeting and I think it relates to where the molecular biologists are beginning to see highly specific tissue targeting. It seems that these compounds that you are looking at are indeed moved to specific locations. It has been commented that there are these major differences between the various tertiary and quaternary compounds, that these are products of multienzyme steps, and that some of these enzymes are not soluble. So it seems to me that there must be specific transport mechanisms. I think that there are great opportunities for looking much more carefully at the subcellular events that are occurring. The various intermediary compounds must be passed along a chain to these various enzymes.

My questions are therefore several. For example, Professor Yamada mentioned that he believes there is at least one step at which one enzyme catalyses two reactions. Are there many instances where these enzymes may catalyse multiple steps, for example, in these nine and fifteen step chains? Presumably these insoluble fractions are on membranes; how much is known about this, beyond what we have heard today? Can we be enlightened on the multiplicity of steps of any of these enzymes and on the actual targeting and migration?

Zenk: One of the systems which I tried to describe was the vesicular system where the enzymes are within these vesicles in a soluble form. If you freeze the vesicles and destroy the membrane, the enzymes are released from the vesicles.

Hall: Are they as functional when they are soluble as when they are attached to a membrane?

Zenk: We have to postulate that these transport proteins are in the tonoplast. They are definitely membrane bound, for example, carrier systems or ATPases. There is very little information on the carriers or transporters for secondary metabolites. There is more information on sucrose transport, the carriers for which are being studied by A. Maretzki's group in Hawaii.

Hall: Even though the enzymes may have the same activity when released *in vitro*, in the cell presumably they are not released, they are held and probably targeted to specific membranes and dispositions. Therefore these compounds presumably must be passed along in some way.

Zenk: Yes, in the secondary product field we know nothing about the transport of these enzymes to their final destination.

Ohyama: Concerning the genetics of the chloroplast: we found a gene for what we have called the membrane-binding protein. This is a very active nucleotide-binding protein and probably passes between the cytoplasm and the chloroplast. It is involved with a transport system but we don't yet know what the substrate is.

Zenk: Did you try sucrose?

Ohyama: This is probably a sugar transporter but we don't know yet.

Rhodes: Only the last four enzymes of the pathway leading to berberine biosynthesis are actually within the vesicle. Do you know anything about the transport of their immediate precursor into the vesicle? Is there specificity at that point?

Zenk: We have looked at that, it's not ATP dependent as far as we can tell; I think it's mere diffusion.

Yamada: You showed that the first step of benzylisoquinoline alkaloid biosynthesis is the condensation of dopamine and 4-hydroxyphenylacetalde- hyde to give norcoclaurine. Formerly, you suggested that 3,4-dihydroxy- phenylacetaldehyde was condensed with dopamine to give norlaudanosoline. Do you think that this condensation reaction is catalysed by a single enzyme with a low substrate specificity or that two different enzymes are responsible for the reaction?

Zenk: We published details of the enzyme some time ago (Schumacher et al 1983). It exists as four isoenzymes, and these accept three phenylacetaldehyde substrates, namely 3,4-dihydroxy aldehydes, the 4-monohydroxy and the un- hydroxylated aldehydes. The unhydroxylated (C-ring) alkaloid is a completely artificial compound. This first enzyme seems therefore to be unspecific towards the aldehyde component, but it is very specific for the amine component.

Yamada: I think that there are two kinds of aldehydes in the cells but the intracellular location of the condensation enzyme leads to the production of norcoclaurine at some times and norlaudansoline at other times.

Zenk: That is not what we have found recently. I think the enzyme exclusive- ly works with a monohydroxylated aldehyde. The dihydroxylated aldehyde is not the natural precursor *in vivo*. We also found an enzyme which decarboxy-lates parahydroxyphenyl pyruvic acid to the monohydroxyaldehyde (Rueffer & Zenk 1987). This then condenses stereospecifically with the amine.

Barz: I would like to ask you about the vacuole of these alkaloid-producing cells. The uptake mechanism seems to differentiate between tertiary and quaternary compounds and between the R and S forms. From that point of view, it's highly specific. You also said that it can exclude nicotine. Can this be generalized to say that these vacuoles will only take up endogenous alkaloids and no other compounds, such as organic acids or glucosides of different types? I thought that plant cells take up everything that is presented to them and that all of these compounds are finally transported to the vacuole. Are these exclusively alkaloid vacuoles?

Zenk: No, the substances we are talking about are all alkaloidal in nature. We did not talk about primary metabolites, such as glucosides or organic acids. I think we have to stick to the alkaloid. I would like to make the point that these vacuoles are so specific that they only take up those alkaloids which are indigenous to the plant species. An isoquinoline-containing plant takes up only isoquinolines, it discriminates against tropane alkaloids, indole alkaloids and other alkaloids. An indole alkaloid-containing plant takes up only indole alkaloids and does not take up isoquinolines, for instance. We published this (Deus-Neumann & Zenk 1984, 1986) and the results were corroborated by other laboratories.

References

Deus-Neumann B, Zenk MH 1984 A highly selective alkaloid uptake system in vacuoles of higher plants. Planta 162:250–260

Deus-Neumann B, Zenk MH 1986 Accumulation of alkaloids in plant vacuoles does not involve an ion-trap mechanism. Planta 167:44–53

Mende P, Wink M 1987 Uptake of the quinolizidine alkaloid lupanine by protoplasts and isolated vacuoles of suspension-cultured *Lupinus polyphyllus* cells. Diffusion or carrier-related transport. J Plant Physiol 129:229–243

Rueffer M, Zenk MH 1987 Distant precursors of benzylisoquinoline alkaloids and their enzymatic formation. Z Naturfo C 42:319–332

Schumacher HM, Rueffer M, Nagakura N, Zenk MH 1983 Partial purification and properties of (S)-Norlaudanosoline synthase from *Eschscholtzia tenuifolia* cell-cultures. Planta Med 48:212–220

Industrial production of shikonin and berberine

Yasuhiro Fujita

Bioscience Research Laboratories, Mitsui Petrochemical Industries, Waki-Cho, Kuga-Gun, Yamaguchi-Ken, 740 Japan

Abstract. We have established industrial production processes for shikonin and berberine, secondary plant metabolites that have commercial and pharmaceutical uses, and have been producing shikonin commercially since 1983. Enhanced production of target metabolites is essential in order to lower the high production costs involved in the commercial process. Basic ways to increase productivity, thereby lowering costs, are 1) acquisition of a highly productive cell line, and 2) establishment of culture conditions and procedures that ensure maximum productivity of the cell line.

We obtained a highly productive berberine-producing cell line of *Coptis japonica* by repeated cell aggregate selection, and a highly productive shikonin-producing *Lithospermum erythrorhizon* cell line by protoplast selection. Then a two-stage culture method was established and the optimal concentrations of the components of the medium used at each stage were determined for the production of shikonin. Our berberine yields also were enhanced by the development of a method in which *Coptis* cells are cultured continuously at a density five times that used ordinarily.

1988 Applications of plant cell and tissue culture. Wiley, Chichester (Ciba Foundation Symposium 137) p 228–238

Research on the production of useful secondary metabolites by plant cell cultures has been going on for the past forty years, during which time the related technology has shown remarkable advances, resulting in our large scale commercial production of shikonin by cell cultures of *Lithospermum erythrorhizon*. However, there has been no industrial-scale production of other secondary metabolites produced by plant cell cultures, in spite of enterprising studies by researchers throughout the world. The main reason for the absence of industrial applications of results from metabolite production research is that production costs of the target compounds are much higher than expected. Therefore it is essential to lower production costs, and the way to do this is by enhancing the production of the target compounds by plant cell cultures.

I here present the results of our studies on the enhancement of shikonin and berberine production with emphasis on the establishment of highly productive cell lines and on optimal culture conditions and procedures.

Acquisition of a highly productive cell line

Highly productive cell lines usually have been obtained by selection; when the parent cell line has low productivity, this method is very effective. We obtained cell lines that gave high yields of berberine by the selection of *Coptis japonica* cell lines which had been derived from small cell aggregates of a parent cell line. By repeating this selection procedure, a *C. japonica* cell line that was five times more productive than the original line was obtained. The stability of the productivity of selected cell lines has also been studied in our laboratory. Productivity has fluctuated to some extent but generally production has been relatively stable for several years. We have continued to repeat selection but there has been no further enhancement of productivity by this method.

We analysed the berberine content of individual cells from high yield cell lines by flow cytometry to determine why lines producing much higher yields cannot be obtained by colony selection. We found that the individual cells of a selected cell line vary greatly in their intracellular berberine content. This may be because lines selected by this procedure are derived from aggregates that may have contained cells with heterogeneous metabolite productivity. Therefore, we have investigated protoplast isolation by cell sorting and the culture of only cells with high berberine content. Isolation of highly productive individual cells using a cell sorter proved to be possible. Many calli could be derived from the protoplasts thus isolated, and we are now examining the berberine productivity of these callus cells.

When the target metabolite has anti-bacterial activity, as does berberine, and it can be secreted from the cell, the efficiency of selection by the colony method can be enhanced by combining it with a bioassay. We have used such a combined procedure to obtain lines that are highly productive for berberine, e.g. *Thalictrum minus* cell culture (Suzuki et al 1987).

TABLE 1   Shikonin productivity of selected cell lines

Cell line	Growth rate (g/g inoculum)	Shikonin content (%)	Shikonin productivity (g/g inoculum)
parent	23.9	17.6	4.20
A	27.8	23.2	6.45
B	26.4	22.6	5.97
C	31.2	18.5	5.78
D	29.1	19.5	5.68
E	25.2	21.7	5.47

Small colonies picked from the protoplast culture plate were cultured on Linsmaier-Skoog agar medium (1965) to establish subclones. The growth rate and the shikonin production of each subclone were investigated in a two-stage culture (nine days in the 1st stage culture for the cell growth and 14 days in the 2nd stage culture for shikonin production). The growth rate was calculated from the growth rates obtained for the 1st and 2nd cultures. Shikonin content was measured in cells harvested after 20 days of culture.

Single-cell selection using protoplasts is a good way to obtain cell lines that are stable and highly productive for target metabolites. We recently reported the selection of high yield lines for shikonin derivatives from a protoplast culture of *L. erythrorhizon* cells (Fujita et al 1985). When we examined the shikonin production of our protoplast-derived cell lines, we found that their productivities were distributed widely and equally around the corresponding value for the parent line. The average productivity was similar to that of the parent cells. When a highly productive cell line (4.2 g shikonin/g inoculum/23 days) was the parent, a superior daughter line with 1.5 times higher productivity (6.5 g shikonin/g inoculum/23 days) could be obtained by protoplast selection (Table 1). Moreover, the productivity of this protoplast-derived line was more stable during subculture than that of its parent line, which had been obtained by the usual selection method.

Selection is effective for obtaining highly productive cell lines for secondary metabolite production, but it is not possible to obtain a cell line better than the original line derived from the most highly producing cell in the parental cell aggregate. Therefore new ways must be found to obtain truly superior cell lines.

Plant cell fusion is one of the most promising new methods for the establishment of a cell line that is truly highly productive for a target metabolite. We have met the challenge of obtaining lines that produce high yields of shikonin derivatives by fusing high metabolite content and high growth cells of *L. erythrorhizon*. We first established an efficient fusion method in which the fusion frequency was greatly increased by controlling the osmotic pressure (Takahashi et al 1986). We then developed a selective isolation method for fused protoplasts using a cell sorter. Having obtained many calli from our fused protoplasts isolated by this method, we are now examining their shikonin productivities.

Optimization of culture conditions

Optimizing the concentrations of the components in a medium is generally an effective way to improve the production of secondary metabolites, as well as to select a highly productive cell line. Development of a new culture medium markedly increased metabolite production by *L. erythrorhizon* cell cultures. *L. erythrorhizon* cells do not produce shikonin derivatives in suspension cultures, but Tabata et al (1974) were able to produce shikonin from *L. erythrorhizon* callus cultures. Our search for the best medium for shikonin production showed that ammonium ions strongly inhibit shikonin production in suspension cultures. Although no shikonin was produced in Linsmaier-Skoog (1965), Gamborg's B-5 (1968), or Nitsch and Nitsch's (1969) media, shikonin production did take place in White's medium (1954) that contained only nitrate as a nitrogen source (Fujita et al 1981a).

TABLE 2   Effect of a two-stage culture on shikonin production

	One-stage culture W[a]	Two-stage culture LS[b]–W	MG-5–M-9
Culture period (days)	14	23	23
Yield of shikonin derivatives (mg)	38	290	3670
Ratio of yields from the different media	1	5	58

[a] White's medium (1954); [b] Linsmaier-Skoog medium (1965). In the two-stage culture, cells resulting from the first stage were transferred to the production medium for the second stage. The media used were LS medium, W medium and our newly developed media, MG-5 (for cell growth) and M-9 (for shikonin production).

L. erythrorhizon cells, however, did not grow well in White's medium, whereas they grew vigorously in Linsmaier-Skoog medium. Therefore we devised a two-stage culture method in which cells were cultured for proliferation in Linsmaier-Skoog medium (first stage) then, using White's medium which was suitable for shikonin production (second stage), we obtained shikonin derivatives from the second stage cells. Furthermore, we optimized the concentration of each compound in the media used for both cell growth (MG-5) and shikonin production (M-9) (Yamada & Fujita 1984, Fujita et al 1981b, 1982). Thereby the production of shikonin was enhanced in comparison to production using Linsmaier-Skoog and White's media (Table 2).

Culture procedures

Increasing the production of the target metabolite per unit volume of culture tank is the most important factor for its commercial production. An increase in cell density is the most direct way to do this, but the maximum cell density varies with the species and culture conditions used because the water contents of plant cells differ greatly.

When our *C. japonica* cells settled in the production medium, their packed cell density was about 100 grams dry weight/l, which was too dense for culture. When the density was 70 gDW/l, culture was possible in the commonly used aerated, agitation-type jar fermentor.

Engineering problems, such as how to agitate cells sufficiently without destroying them and how to maintain an adequate supply of oxygen, must be solved in order to culture cells at a high cell density. The critical element is the supply of nutrients: as the number of cells increases, a greater supply of nutrients is required, but when the concentrations of nutrients become very high, cell growth and the production of berberine are prevented. As shown in Fig. 1, when we increased the concentration of each component in the

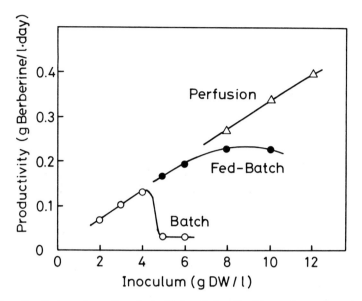

FIG. 1. *Coptis japonica* cultured at a high cell density. The optimum inoculum in the batch culture with modified Linsmaier-Skoog (LS) medium was 2.0 g dry weight/l. For the batch culture, medium in which the concentrations of all the components had been increased in proportion to the inoculum size was used. In the fed-batch and perfusion cultures, modified LS medium was used at the beginning of culture, nutrients being added according to cell growth during the culture period. In the perfusion culture, the volume of the medium was constant, with part being continuously replaced by fresh medium. The concentration of dissolved oxygen in the medium was maintained at 12 parts per million in both cultures.

medium in proportion to the increase in the size of the inoculum, in the batch culture of *C. japonica* cells the cell yield increased proportionally but only up to 4 gDW/l of inoculum.

One way to overcome this limitation is to use the 'fed-batch culture.' As this method does not raise the concentration of individual nutrients, inhibition of cell growth and of berberine production is largely prevented. But even in the fed-batch culture, berberine production was gradually inhibited at an inoculum size greater than 6 gDW/l (Fig. 1). In such high density cultures, secretions (such as superannuated cell substances) often function as inhibitors of cell growth and berberine production. To overcome this handicap, we developed a perfusion culture method in which part of the medium is renewed continuously, thereby forestalling the accumulation of harmful substances. Using this method, we could culture *C. japonica* cells at a density five times the ordinary density used, and obtained a 70 gDW/l cell yield with no decrease in berberine content.

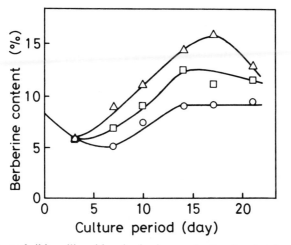

FIG. 2.   Effect of gibberellic acid on berberine production in a batch culture of *Coptis japonica*. Gibberellic acid (GA_3) was added to the medium at the beginning of culture: ○, control; □, 10^{-8} M; △, 10^{-5} M.

Continuous culture is another method by which to enhance the yields of target compounds. We established a form of continuous culture that involves a perfusion technique for berberine production at a high cell density. Our continuous culture has the advantage that cells can be cultured when the growth rate is at its maximum. However, conditions suitable for cell growth are not suitable for berberine production because, as with most other plant metabolites, production occurs a few days after cell proliferation. Therefore we developed two methods: first, we obtained efficient production of berberine when the cell growth rate is maximal by reducing the time lag between cell growth and berberine production. Second, because the conditions for the separate stages of cell growth and berberine production are optimal, cells

TABLE 3   Effect of gibberellic acid in a high density continuous culture of *Coptis japonica* cells

	Concentration of gibberellic acid	
	0	*10^{-7} M*
Specific growth rate (l/h)	0.21	0.17
Berberine content (%)	1.5	5.0
Berberine production (g/l/day)	0.22	0.60

The cell density of the culture was maintained at 70 g dry weight/l. Berberine production was stable for more than two months.

proliferate vigorously in the first-stage culture tank, then produce berberine at the maximum rate in the second-stage tank.

We found that addition of gibberellic acid to our batch culture of *C. japonica* shortened the time lag between cell growth and berberine production (Fig. 2) (Morimoto et al 1986) and enhanced berberine production in our high cell density, continuous culture (Table 3). We are now studying this enhancement using our two-stage continuous culture.

Conclusions

We have obtained cell lines that show extremely high productivities for shikonin and for berberine and have established the optimal culture conditions and procedures needed for each cell line. There are many instances, however, in which cells do not produce the intended compound, or produce very little of it. For such cells we must find how to induce metabolite production and how to enhance the cell line's productivity. The techniques we have established for shikonin and berberine production may provide the bases for doing this.

Acknowledgements

I am deeply grateful to Professor Y. Yamada, Kyoto University, for his generous gift of the *C. japonica* cell lines and his continuing help, and to Professor M. Tabata, Kyoto University, for his gift of the *L. erythrorhizon* cell lines and useful advice. Thanks also are due to Professor A. Nishi, Toyama Medical and Pharmaceutical University, for his beneficial suggestions about our research.

References

Fujita Y, Hara Y, Ogino T, Suga C 1981a Production of shikonin derivatives by cell suspension cultures of *Lithospermum erythrorhizon*. I. Effect of nitrogen sources on the production of shikonin derivatives. Plant Cell Rep 1:59–60

Fujita Y, Hara Y, Suga C, Morimoto T 1981b Production of shikonin derivatives by cell suspension cultures of *Lithospermum erythrorhizon*. II. A new medium for the production of shikonin derivatives. Plant Cell Rep 1:61–63

Fujita Y, Tabata Y, Nishi A, Yamada Y 1982 New medium and production of secondary compounds with the two-stage culture method. In: Fujiwara A (ed) Plant tissue culture 1982, Maruzen, Tokyo (Proc 5th Int Congr Plant Tissue & Cell Culture, Tokyo and Lake Yamanaka, 1982) p 399–400

Fujita Y, Takahashi S, Yamada Y 1985 Selection of cell lines with high productivity of shikonin derivatives by protoplast culture of *Lithospermum erythrorhizon*. Agric Biol Chem 49:1755–1759

Gamborg OL, Miller RA, Ojima K 1968 Nutrient requirements of suspension cultures of soybean root cells. Exp Cell Res 50:151–158

Linsmaier EF, Skoog F 1965 Organic growth factor requirements of tobacco tissue cultures. Physiol Plant 18:100–127

Morimoto T, Hara Y, Yoshioka T, Fujita Y, Yamada Y 1986 Production of berberine by suspension cultures of *Coptis japonica*. In: Somers DA et al (eds) Abstr 6th Int Congr Plant Tissue & Cell Culture, International Association of Plant Tissue Culture, Minneapolis, 1986, p 68

Nitsch JP, Nitsch C 1969 Haploid plants from pollen grains. Science 163:85–87

Suzuki T, Yoshioka T, Hara Y, Tabata M, Fujita Y 1987 A new bioassay system for screening high berberine-producing cell colonies of *Thalictrum minus*. In: Neijssel OM et al (eds) Proc 4th Eur Congr Biotechnol, Amsterdam, 1987. Elsevier Science Publishers, Amsterdam, vol 2:380–383(abstr)

Tabata M, Mizukami H, Hiraoka N, Konoshima M 1974 Pigment formation in callus cultures of *Lithospermum erythrorhizon*. Phytochemistry (Oxf) 13:927–937

Takahashi S, Fujita Y, Yamada Y 1986 Effective method for protoplast fusion. In: Somers DA et al (eds) Abstr 6th Int Congr Plant Tissue & Cell Culture, International Association for Plant Tissue Culture, Minneapolis, 1986 p 317

White PR 1954 The cultivation of animal and plant cells. Ronald Press, New York

Yamada Y, Fujita Y 1984 Production of useful compounds in culture. In: Evance DA et al (eds) Handbook of plant cell culture, Macmillan, New York p 717–728

DISCUSSION[1]

Fowler: I would like to ask two technical questions—what are your stirring rates in the fermentor and what are the gassing rates?

Fujita: I don't remember the values at present.

Fowler: Perhaps I should explain the reason for the question. You are working at 70 grams/l dry weight, which is a very high cell density. One would have thought that mixing at such a high density to achieve homogeneity must be very difficult. You presumably have to provide a large power input to the system to get adequate mixing and I wonder how much damage is being done to the cells in consequence.

The second part of the question relates to indications from our own work and that of others that the gassing regime, both in composition and in the rate at which it is applied, can greatly affect cell growth rate and product synthesis. I would be interested to hear your experience with these parameters.

Fujita: We tested the agitation effect for *Coptis japonica* cells. When the cell concentration exceeds 80 g/l, the viscosity goes up rapidly, but under 75 g/l we could get smooth agitation and we observed no harmful effects on the cells.

For the aeration we controlled the concentration of dissolved oxygen using air and pure oxygen gas.

[1] Dr Fujita's comments were translated by his colleague Dr Yoshioka.

Fowler: So you took air and then put what I would call a 'bleed' system of pure oxygen into the tank, as well?

Fujita: We add the air from the bottom of the tank at the constant gassing rate and with the same nozzle we feed regulated pure oxygen, and we control the concentration of dissolved oxygen.

Scowcroft: An observation that a plant breeder might make about the work that has been discussed today, on using cell cultures to produce a high level of berberine, concerns the degree of genetic diversity that can be explored. For example, Dr Fujita has described, both with berberine and shikonin, that they started from an existing cell line and selected for increased levels of a particular compound. Despite the possible influence of somaclonal variation, it seems to me that you are starting from an extremely narrow genetic base. My comment is: now that you know you can do it, go back to the germplasm diversity of a particular species, such as *C. japonica* for berberine, and see if you can derive new cell cultures from different races of that species to achieve higher levels. If you have a narrow genetic base,there is a limit as to how far you can get.

Galun: We have had this discussion before, some 5–6 years ago.

Scowcroft: You may well have done so, but I haven't seen any evidence that if you go back to the germplasm base of the species concerned, higher yielding cell lines cannot be obtained.

Cocking: I should like to point out that colleagues at Mitsui Petrochemical Industries Limited have undertaken some genetic manipulation of their cell cultures by what they have been doing with protoplast fusion, and have been trying to by-pass this baseline level of performance which is inherent in the cultures. We have seen a rare example—and I had the privilege of visiting Dr Fujita and his colleagues at Mitsui just before this meeting—of workers in the plant secondary product field using some of the genetic manipulation methodology that is now available. It was very exciting to see the use of a fluorescence-activated cell sorter in this connection. If I could ask a more general question, put on my botanical hat and refer to the plants from which shikonin and berberine are derived. Historically, in Japan *Lithospermum erythrorhizon* was used, and is perhaps still used in normal horticultural and arable practice to produce shikonin. Is there any comparison of the economics of production by the use of this fermentation process with the direct isolation of shikonin from naturally grown plant *Lithospermum* material? I think that would be very interesting to know and I do hope that some cost figures are available.

Fujita: For the shikonin production, the cost of production by fermentation is lower than the extraction from intact plants. We believe that the fermentation process is advantageous.

Yamada: Roughly how much lower?

Fujita: We cannot say that because that would affect the market price!

Yamada: *L. erythrorhizon* does not grow well in Japan now because of air pollution. Therefore it is very difficult to compare prices.

Cocking: That was really part of my question, Chairman. Of the total production of shikonin in Japan during the last five years, what percentage of the production has been by this process as compared with natural production?

Fujita: The market for shikonin from naturally grown plants was small so we have developed new usages for the fermentation products. We cannot say how much of the total production is by the fermentation method.

Yamada: In Japan, a lot of shikonin was previously used for dyeing and topical medicine. The amount of shikonin available was then reduced as a consequence of environmental pollution, so Mitsui company wanted to produce it by a tissue culture method.

Cocking: A percentage please, Chairman.

Yamada: Because of the effects of pollution on natural production, that figure cannot be calculated.

Zenk: I think all the values that were shown for production were expressed as berberine. I have shown a slide of a chromatogram of a *Coptis* strain which contains considerable amounts of other protoberberines, apart from berberine, such as palmatine, coptisine itself and jatrorrhizine. I wonder whether these figures which you showed, Dr Fujita, have been corrected to account for this. Do they represent total protoberberines or true berberine?

My second point is that since these protoberberines probably are in the *Coptis* strain, if you want to end up with pure berberine you have to undertake a major purification procedure, which will be very costly.

Fujita: Our *Coptis* cell lines also produce coptisine and palmatine. The value we gave is the pure berberine content. The recovery process is rather complicated, but we have developed a purification process that is relatively inexpensive, that is not such a big problem.

Zenk: I think this also answers Dr Scowcroft's question. In my estimation, you would have about the same amount of berberine as you have of the other protoberberine side products. So if you have about 7 g of berberine then you have about 14–15 grams of total protoberberines. I think this is about the limit of what plant cells can produce and hold, so I don't think there is any point in looking for more genetic variation.

Scowcroft: A relevant point concerns breeding and physiology. About ten years ago plant physiologists estimated that the maximum yield potential of wheat was about 12 tonnes per hectare. Now farmers achieve fifteen tonnes per hectare in the UK. You say it may not be possible to enhance production of secondary metabolites by utilizing germplasm diversity, but where is the evidence that you have screened the genetic diversity for berberine, for shikonin? Professor Zenk, that was the second point of your proposed strategy, namely selection of high-yielding plants. I don't work in this area so I ask that question, why not take account of the genetic diversity?

Zenk: I think this is a relatively simple sort of thing. Your figures for wheat yield rely on photosynthesis. In secondary metabolite production there is a limited sucrose supply. You can only convert so much sucrose into a secondary product, a cell can accumulate only about 40% of its dry weight as secondary products.

Riley: Can I ask Dr Fujita some questions which are entirely those of the layman in this field. I didn't see where the nitrogen costs and the energy costs were incorporated in the first diagram. Are they important?

Fujita: The nitrate cost is negligible.

Riley: At what temperature do you run cultures to produce berberine and shikonin? Also in cell sorting, how do you select for heterokaryons against homokaryons?

Fujita: The temperature is 25°C in each case. To sort the heterokaryons, we use different coloured dyes for the two cell lines. When the cells fuse, the colour changes and this is detected by the cell sorter.

Problems in commercial exploitation of plant cell cultures

M.W. Fowler

Wolfson Institute of Biotechnology, The University of Sheffield, Sheffield S10 2TN, UK

Abstract. With the commercialization of a number of processes in the area of plant cell biotechnology and natural product synthesis, and growing, though sometimes highly sceptical, interest by industry, attention has begun to focus on those areas which still cause major problems on the path to commercial application. While there is no question that the plant kingdom, with a tremendously diverse and flexible gene pool, has the potential to provide us with what we need, do we have the understanding and means of harnessing that potential in a commercial framework? This paper considers some of the key areas which currently limit a more widespread industrial application of plant cell culture technology. These areas include enhancement of growth rate, product synthesis and productivity, the performance of cultures in different forms of process plant, the stability of cell lines and, often forgotten but crucially important, the mode of product recovery. Underlying all this, however, is the constantly recurring theme that the real limitation lies in our lack of understanding of the basic physiology and biochemistry of plant cell systems. Until we begin to build a fundamental knowledge base in a more constructive fashion than at present, progress towards application will be restricted and will all too often be based upon serendipitous observation rather than scientific logic.

1988 Applications of plant cell and tissue culture. Wiley, Chichester (Ciba Foundation Symposium 137) p 239–253

Rapid developments in plant biotechnology in recent years have provoked much discussion about the overall potential of commercial application of the technology. Specifically, in the area of natural product synthesis a number of systems have been developed commercially using cell and tissue culture as the process vehicle. Despite these successes, many parts of industry are still sceptical about the potential of plant cell culture, pointing to difficulties over product yield, growth rate, problems of expression of desired traits, and raising questions about the complexity of process operation and its capital cost. Undoubtedly, the plant kingdom has a tremendous versatility and repertoire in natural product synthesis and enzymology and there is an extensive literature in support of this. The key question arises as to how we can harness that potential in a way which makes cell culture processes economically viable and competitive in the market place.

The space available does not allow a detailed consideration of each facet of the development of a plant cell culture process but the key areas, with possible new approaches, are highlighted. The ultimate criterion of success for any cell culture process is its commercial viability, either in competition with alternative technologies or in presenting new products at a price which the consumer is prepared to pay. The term 'commercial viability' covers a number of elements, some economic, some technical: this presentation is concerned with the latter. From a technical standpoint, the key element in commercial viability has to be 'productivity', measured in terms of amount of product synthesized per unit working volume of the reactor in a given period of time. The time element in this calculation is often forgotten by scientists but is crucially important for the process plant manager. He needs to have the reaction system operating at maximum productivity for the greatest period of time during any production run to justify the large capital cost of the process plant investment. It may be economically more advantageous to have a culture system which has a slightly lower level of productivity maintained over

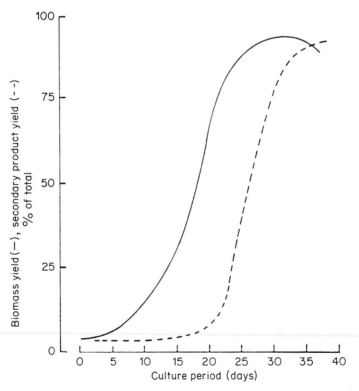

FIG. 1. Growth-dissociated secondary product accumulation as typically seen in plant cell cultures. The culture period is indicative only, but is nonetheless typical of many cell suspension systems.

TABLE 1 **Effect of sucrose concentrations on growth and product formation by** *Catharanthus roseus* **ID1**

	Sucrose concentration (%)			
	2	*4*	*6*	*8*
Maximum wet weight (g l^{-1})	278.2	372.0	506.0	464.0
Maximum dry weight (g l^{-1})	13.0	25.0	31.0	39.0
Maximum serpentine (mg g dry wt^{-1})	5.24	9.02	9.24	8.05
Serpentine yield (mg l^{-1})	44.7	145.2	149.7	163.0
Serpentine productivity (mg l^{-1} day^{-1})	2.79	7.26	5.35	6.27

an extended period of time than one with a very high level of productivity which is achieved only after a considerable time lag and then maintained only for a short period. Unfortunately, the literature describes rather a large number of examples of the latter situation. Various factors which affect the productivity of plant cell/tissue culture systems are discussed below.

Growth rate and biomass yield

Over the years, a great deal of attention has been paid to cell growth rate. A key point is that no matter what sort of process format is adopted, there is always an initial need to develop a large amount of plant cell material as quickly as possible. Unfortunately, in most cases natural product (secondary metabolite) synthesis is uncoupled from cell growth and division (see Fig. 1), therefore those factors which enhance growth rate tend to lower product yields.

A number of factors have been found to affect growth rate and also biomass yield. Nutrient medium composition is an obvious one, the level and nature of the nitrogen source and carbon source being of key importance.

TABLE 2 **The effect of temperature on growth and product formation by** *Catharanthus roseus* **ID1**

	Temperature		
	20 °C	*25 °C*	*30 °C*
Maximum wet weight (g l^{-1})	169.2	284.0	238.0
Maximum dry weight (g l^{-1})	10.9	13.0	11.9
Growth rate (day^{-1})	0.19	0.41	0.69
Doubling time (days)	3.7	1.7	1.0
Maximum serpentine (mg g dry wt^{-1})	2.26	2.70	0.12
Serpentine yield (mg l^{-1})	20.7	44.7	0.87
Serpentine productivity (mg g^{-1} day^{-1})	1.15	2.79	0.07

TABLE 3   The maximum exponential growth rates and corresponding doubling times for *C. roseus* suspension cultures grown in a seven litre air-lift bioreactor operated at a range of aeration rates

Aeration rates ($dm^3\ min^{-1}$)	Mean doubling time (hours)	Exponential growth rate (day^{-1})
2	39 ± 2	0.42
3	46 ± 2	0.36
4	58 ± 17	0.29
6	72 ± 12	0.23

Early work by Noguchi and colleagues (Noguchi et al 1977) demonstrated a close relationship between the rate of growth of tobacco cell cultures and the level of nitrogen supply. Work in our own laboratories on carbon supply has shown a similar relationship with the level of sucrose supplied to a culture. The carbon source has an additional effect on biomass yield; we have observed with a number of cell lines an almost linear relationship between carbon supply and biomass yield up to between 30–50 gram per litre dry weight biomass produced. An example of this is shown in Table 1 from our work with *Catharanthus roseus* cell cultures. In simplistic terms, one might argue that the higher the biomass level per litre of culture, the greater the catalytic potential, but there is a limit to the level of biomass loading which can be handled effectively in a conventional bioreactor (see below).

Temperature and its effect on culture performance has been somewhat neglected in studies with cell cultures. In a recent, fairly comprehensive survey of some of our own cell lines, two points have emerged; first that temperature has some quite dramatic effects on features such as growth rate, biomass yield and productivity, and second that the nature of the effect varies from cell line to cell line. An example of the type of observation we have made with our own *Catharanthus* cell lines is shown in Table 2. It is clear from these data that there are situations where we need to pay closer attention to temperature monitoring and control in cultures than previously thought.

A factor influencing growth rate and biomass yield which has been comparatively little investigated but which could markedly affect process performance is that of the gassing regime in terms of oxygen and carbon dioxide supply. (Other gases, such as ethylene, may also have effects but for perhaps rather different reasons.) Work in our own (Hegarty et al 1986) and one or two other laboratories has shown that the composition and rate of gas supply can profoundly affect a culture in a variety of ways. The problem is particularly acute with air-driven bioreactors, such as draught tube or loop reactors, because of the high venting rates required to provide adequate mixing of nutrient and biomass. Table 3 contains data relating to the effect of ventilation rate on doubling time and exponential growth rate in cultures of *C. roseus* grown in a seven litre air-lift loop reactor. Increasing ventilation rates by quite modest amounts compared with microbial cultures caused a major reduction in exponential growth rate and a concomitant increase in culture doubling time. At the higher ventilation rates the onset of the period of maximum exponential growth was also delayed (data not shown). At the other end of the scale, with ventilation rates of around 2 ml l^{-1}, maximum growth rates were achieved but cultures soon became oxygen limited at a dissolved oxygen concentration of about 20%. So while an initially high growth rate appears to be best achieved at low ventilation rates, oxygen supplementation is probably required to maintain high growth rates throughout the culture period.

The 'knock-on' effects of high gassing or venting rates on the physiological state and the performance of a culture can be marked and, moreover, of a quite different nature to anything seen with microbial cultures grown in air-driven reactors. We have noted two particular effects in our own work. First, at high venting rates there is a loss or 'stripping off' of carbon dioxide, which leads to lowered growth rate and biomass yield (an indication of this for growth rate is seen in Table 3). Second, plant cell cultures do not appear to have the very high oxygen demands seen for microbial cultures (plant cells typically require 1–2 mM O_2 per hour; microbial cells operate in a much wider range of 10–50 mM O_2 per hour) and their performance may even be adversely affected by high levels of dissolved oxygen. The problem of loss of carbon dioxide can be coped with by providing a bleed system in the bioreactor to maintain the level of dissolved carbon dioxide. Although the plant cells in our studies are not photosynthetic, there does seem to be a key biosynthetic requirement for carbon dioxide (at least as bicarbonate) supplied through anaplerotic reactions. Such is the tight control and balance over this in rapidly growing cells that a small loss of carbon dioxide caused by a high rate of ventilation does restrict metabolism, and hence growth.

The situation with oxygen is far more problematical. Our current view is that a high oxygen concentration may result in a slowing of growth and cell division, through inhibition of certain key steps in the central pathways of respiratory metabolism. This would lead to reduced culture performance manifested in a number of ways. Unfortunately, this has to be balanced against problems of oxygen limitation where there is a need to maintain high growth rate (see above). Our own studies with gassing regimes in large scale vessels indicate that there are complex interactions here which could profoundly affect our approaches to process control in the development of an industrial system. This is an important area for future process development work. It also raises the question as to whether we have sufficiently reliable gas monitoring and control systems to cope with the extended run times experienced with plant cell cultures.

Product yield

The literature is full of learned discussion about possible approaches to enhancement of product yield in plant cell culture, an aspect of the technology which has always been seen as the limiting area in commercial application. There appears to be no one strategy that is generally applicable with plant cell cultures to all product targets; indeed, given the range of systems and products with which we are concerned, it would be surprising if there were. The key point from the process development point of view is that a sufficient range of options is available to make success a strong possibility. I have avoided the temptation to provide yet another table of examples of cell cultures which give product yields equivalent to or greater than the parent plant (see Berlin 1986, Stafford et al 1986). The progress in developing high performing cell and tissue culture lines in recent years can only be welcomed, as can work on the effect of elicitor systems in inducing high product yields, together with the more recent observations that so-called 'hairy root' cultures not only provide high product yields but also maintain high culture growth rates (Hamill et al 1987). I think it important, however, that each of these systems is considered in perspective. There are examples, for instance, where elicitation and hairy roots have been found not only to have no effect on product synthesis, but to actually reduce it. This is certainly not a firm foundation upon which to begin building process models and systems. We urgently need a more fundamental understanding of the biochemical mechanisms which underly phenomena such as elicitation and hairy root formation to achieve a more directed approach than at present.

Another key concern in product synthesis which profoundly affects commercial application is culture stability. There are indications that the more selected a cell line is, the more unstable it becomes in relation to product synthesis. However, in the cell lines used in the Mitsui shikonin process (Fujita & Tabata 1987) and in the Natterman process (Ulbrich et al 1985) for rosmarinic acid comparatively high levels of stability have been achieved. We have collected data on several of our cell lines in a rigorous fashion over a number of years and the type of problem that can be encountered is shown in Fig. 2. These data relate to a culture of C. roseus which we have maintained under strict conditions for eight years. This 'instability' may be useful as a selection system, capitalizing upon somaclonal variation. However, once achieved, it has to be brought under control in a process format where large, unpredictable swings in culture performance cannot be tolerated. One must also distinguish between 'stability' in the performance of a cell line and 'variability'. We have recently carried out a complex series of experiments with cell lines of Papaver somniferum, selected for their ability to synthesize not morphine and codeine but other related alkaloids. In this work, we have maintained a basal cell line under a strict subculture regime, paying particular

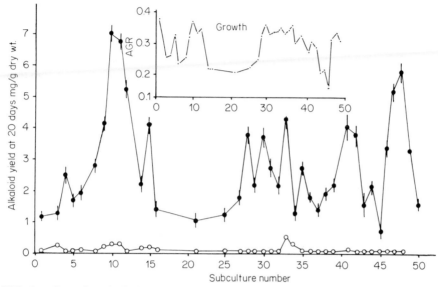

FIG. 2. Growth and alkaloid production of *Catharanthus roseus* cell line C87 during a two year period of serial subculture on M3 medium. Cells were subcultured at 14 day intervals at a dilution of 1:5. Growth and alkaloid content of 20-day-old cells. AGR, average growth rate mg/l/day dry wt accumulation (—·—); serpentine (o), ajmalicine (●). (Morris 1986.)

attention to the technical aspects of the subculture regime itself. We have taken aliquots of this base material at regular time intervals and examined stability of performance when the culture volume is scaled up from 250 ml to one litre and 10 litres. This experiment has now been running for several months with interesting results. As long as we maintain a strict subculture regime, growth rate and biomass yield exhibit about a 5% variation. Product yield shows a slightly higher level of variation, but still within the 10% that is probably tolerable in a process context. Slight variation in the subculture routine does, however, result in significant changes in performance of the cell line for varying periods of time. We have not run the experiment long enough to comment on the long term stability of the system but it does illustrate the need to define precisely the source of any variability in culture performance, i.e. does it arise from loose experimental practice or is it due to some inherent feature of the cell line?

Bioreactor configurations

At the heart of any large scale culture system lies the bioreactor. A whole range of different bioreactor configurations in which to grow plant cells have been used, from simple bubble column systems, through air-driven draught tube and loop rigs to classical stirred-tank reactors. More recently, reactors based on the classical Archimedian screw have been shown to be particularly effective as plant cell culture systems (Ulbrich et al 1985).

The choice of an appropriate bioreactor system has been a focus for much discussion in recent years. Early work with plant cells, including the now classical work of Wagner and Vogelmann (1974), indicated that the fragile nature of plant cells made the use of stirred-tank reactors questionable because of the very high shear forces developed and the consequent rapid loss of culture viability and cellular integrity. Many groups, including our own, therefore began investigation of alternative bioreactor conformations with a lower shear characteristic. Air-driven draught tube reactors in general have a much lower shear characteristic than stirred-tank reactors (Scragg & Fowler 1986). Initial work indicated that these reactors had major potential for large scale cell growth and a number of process formats were developed around them. However, a number of recent observations suggest that air-driven rigs may not be as generally applicable as once thought. The key points are as follows:

1. Calculations by Goldstein and colleagues (1980) indicate that for commercial viability, biomass levels far higher than those currently achieved (and achievable?) in air-driven rigs will be required.
2. Above about 25–30 g l^{-1} biomass, adequate mixing of the culture at nominal gassing rates becomes a problem, resulting in large dead zones within the system.
3. Increased aeration to improved mixing may have deleterious effects on culture performance (see above).

Another important point is that a number of workers have reported on the successful growth of plant cells in traditional stirred-tank reactors. In Western Europe in particular there is considerable spare capacity in conventional stirred-tank reactor systems. If some of this could be harnessed, possibly with some modification, to the growth of plant cells, then it would avoid substantial capital investment for new plant. The importance of this is seen in the Mitsui process for shikonin production which utilizes existing fermentation capacity in stirred-tank reactors. The comments passed by Dr Fujita (Fujita & Tabata 1987) at the last IAPTC Symposium, however, show that this is to some extent a compromise, since growth of *Lithospermum* cells in these stirred-tank reactors resulted in lower product yields compared with small flasks, and there is an accompanying increase in cell damage.

Let us approach this problem from a slightly different direction. The calculations by Goldstein and colleagues (1980) were the first to address in a detailed, albeit somewhat theoretical, way the process economics of plant cell culture systems. While based to an extent on extrapolation from microbial systems because of the lack of plant cell data, they did provide some important sign-posts to the type of growth rates, product yields and biomass yields which would have to be attained to achieve commercial viability. A key point in their calculations relates to the level of biomass loading which may be necessary. At their highest, the levels (over 100 g l^{-1} dry weight) are somewhat above those so far achieved with stirred-tank reactors (about 70 g l^{-1}) and way above those achieved with air-driven systems (30–40) g l^{-1}). To really capitalize on developments in the supporting technology, do we need to develop a whole new generation of bioreactor configurations? Possibly so, and indeed there are signs that this is happening. First we have the highly innovative use of the Archimedian screw system by Ulbrich and colleagues at Nattermann and Company for rosmarinic acid production and then, using a rather different approach, the work by Tanaka (1981) in Japan on thin film rotating drum reactors. These latter are important in that they seem to hold out the prospect of very high biomass loadings and also, because of the large surface area, good gas exchange and control. Developments in this area are awaited with great interest.

Downstream recovery

This is an area which has so far featured little in discussions of plant cell biotechnology, the general view being that classical techniques of steam distillation or solvent extraction would suffice. However, work from one or two groups has indicated that plant cell cultures may differ from the parent plant in that the product may be tightly bound into the cell wall and other structures and not as readily released as from whole plants. Pre-digestion with enzymes seems to be one way of dealing with this problem (Table 3), releasing much of the bound material, which may then be extracted by more conventional means (Fowler 1987). Having a very high-yielding culture is of little use, if the large scale extraction process results in major losses of product. Such a situation could transfer a potentially commercially viable process into a disaster area. More thought needs to be given as a matter of urgency to this problem.

Process format

Space does not allow detailed comment on this but two points, both affecting commercial approaches, should be made.
1. There is a great variety of process options available to the process developer (see Fowler 1987).
2. The majority of those process systems which have come on stream commercially, or are imminent, are essentially single or two-stage batch systems along the lines suggested by Professor Zenk (Zenk 1978).
A final general comment is that these systems are also characterized by their

simplicity in concept; industry does not like complex and often expensive process formats!

Socio-legal considerations

Those involved in plant cell biotechnology directed at the food industry in Western countries will have to become increasingly concerned with a current debate which relates to the concepts of 'nature derived' and 'nature identical'. Within the USA and the European Community (EC) there are signs of both consumer resistance to so-called synthetic additives in foods, and pressure for legislative bodies to exert controls over the whole area of food additives and the use of aids in food processing. This is in addition to the already heavy and rigorous controls on the clinical and veterinarian application of pharmaceutical preparations. Some indications of the problems which might lie ahead come from the recent announcement in the USA by the Food and Drugs Administration that only vanilla from the pod of the vanilla plant may be properly described as natural. This puts vanilla derived from plant cell cultures outside the 'natural' description, and poses question marks over many of the other plant cell products currently under investigation with food industry connections.

Another area of concern relates to that of cell culture-derived enzyme systems and whether or not these, when derived from cell cultures, may be used as food processing aids. There is certainly within the EC at the present time a deep concern over the host organism into which plant genes coding for specific food processing enzymes might be cloned. In general, the cloned enzyme will be considered for approval provided that the organism into which it has been cloned is already approved for use in food, for example, the food yeasts or *Streptococcus lacti* but not *Escherichia coli*. However, it should be emphasized that as yet there are no definitive regulations covering this area in the EC. In Japan the situation would appear to be less restricted.

There is no question that certain decisions by the regulatory authorities, attempting to reflect consumer concerns and interests, could have major effects on the future development and commercialization of plant cell biotechnology, in addition to any technical constraints we may perceive.

Summary

The commercialization of plant cell process systems over the last few years by a number of organizations, but principally those in Japan, has done much to further the cause of plant cell biotechnology and is obviously to be applauded. One has to say, however, that a number of key constraints still prevent a more general application of the technology. Perhaps chief amongst these is our limited ability to control the level of product synthesis, which in turn comes back to our very rudimentary understanding of the physiology and biochemistry of plant cells. It is in this area that I firmly believe we need to focus more effort in the future, if we are to achieve the commercial successes to which we all aspire. I also see the legislative threat as something which begins to loom large in the development of the area. Unless funding organizations can see a clear route to application of this technology, particularly in the food area, then lack of support may greatly restrict the pace of development. It may well be that the professional bodies representing those working in plant cell culture should make concerted and forceful representation to the legislative authorities, particularly in the EC and the USA, regarding application of the products of this technology. If this does not occur, then I think we face far more difficulties from this direction than from inadequacies of our knowledge base or technical developments.

Acknowledgements

We wish to acknowledge the support of the Wolfson Foundation and the UK Science and Engineering Research Council in supporting the work from our laboratory.

References

Berlin J 1986 Secondary products from plant cell cultures. In: Rehm H-J, Reid G (eds) Bio/Technology. V C H Verlagsgesellschaft, Weinheim, vol 4:630–658

Fowler MW 1987 Process systems and approaches for large scale plant cell culture. In: Green CE et al (eds) Plant tissue and cell culture. Alan R. Liss, New York, p 459–472

Fujita Y, Tabata M 1987 Secondary metabolites from plant cells — pharmaceutical applications and progress in commercial production. In: Green CE et al (eds) Plant tissue and cell culture. Alan R. Liss, New York, p 169–186

Goldstein WE, Ingle MB, Lasure L 1980 Product cost analysis. In: Staba EJ (ed) Plant tissue culture as a source of biochemicals. CRC Press, Boca Raton, p 191–234

Hamill JD, Parr AJ, Rhodes MJC, Robins RJ, Walton NJ 1987 New routes to plant secondary products. Bio/Technology 5:800–804

Hegarty PK, Smart NJ, Scragg AH, Fowler MW 1986 The aeration of *Catharanthus roseus* (L)G. Don suspension cultures in airlift bioreactors: The inhibitory effect of high aeration rates on culture growth. J Exp Bot 37:1911–1920

Morris P 1986 Long term stability of alkaloid productivity in cell suspension cultures of *Catharanthus roseus*. In: Morris P et al (eds) Secondary metabolism in plant cell cultures. Cambridge University Press, Cambridge, p 256–262

Noguchi M, Matsumoto T, Hirata Y et al 1977 Improvement of growth rates of plant cell cultures. In: Barz W et al (eds) Plant tissue culture and its biotechnological applications. Springer-Verlag, Berlin, p 85–94

Scragg AH, Fowler MW 1986 The mass culture of plant cells. In: Vasil IK (ed) Cell culture and somatic cell genetics of plants. Academic Press, New York, vol 2:103–128

Stafford A, Morris P, Fowler MW 1986 Plant cell biotechnology — a perspective. Enzyme Microb Technol 8:578–587

Tanaka H 1981 Technological problems in cultivation of plant cells at high density. Biotech Bioeng 23:1203–1218

Ulbrich B, Weisner W, Arens H 1985 Large scale production of rosmarinic acid from plant cell cultures of *Coleus blumei* Benth. In: Neuman KH et al (eds) Primary and secondary metabolism of plant cell cultures. Springer-Verlag, Heidelberg, p 293–303

Wagner F, Vogelmann H 1974 Cultivation of plant tissue cultures in bioreactors and formation of secondary metabolites. In: Barz W et al (eds) Plant tissue culture and its biotechnological application. Springer-Verlag, Heidelberg, p 245–252

Zenk MH 1978 The impact of plant tissue culture on industry and agriculture. In: Thorpe TA (ed) Frontiers of plant tissue culture 1978. IAPTC Calgary, The Bookstore, University of Calgary, p 1–13

DISCUSSION

Scowcroft: I have a technical question: you indicated that as a consequence of inadvertent selection the resistance to shear increases. Does this have a negative effect on downstream extraction? Is it harder to extract from that material?

Fowler: To be quite frank we haven't looked at that particular option yet. We have examined various classical extraction procedures with our shear-tolerant cells and have so far encountered no major problems. What is interesting, is that enzyme pre-digests of our tolerant cells give a greater release of total material from the cell, possibly because of the release of bound substances from the cell wall.

Scowcroft: You talked about extraction procedures: the four methods that you mentioned are pretty savage extraction methods which may damage some fairly critical compounds. Are cell cultures amenable to critical point extraction?

Fowler: There has been a great deal of interest in recent years in supercritical CO_2 extraction of natural products, particularly from people in the flavour and aroma industries where stability and volatilization are important factors in the extraction process. I am not aware that this technique has yet been applied to plant cell cultures, but it could be an important option.

Scowcroft: If you think about the normal position of a plant cell, it is usually surrounded by neighbours and it is immobilized. What about developments in immobilized fermentation technology?

Fowler: The problem is the one that Professor Zenk mentioned this morning, the number of examples of cells where the product is released into the medium is very small. Where chemicals have been used to induce 'leakiness' in cells, one of the key questions to be answered is, whether release is an ongoing process from viable cells or only occurs from the breakdown of dead cells. I get the impression that immobilized cell technology, because of product release problems, is almost on 'hold' at the moment.

Scowcroft: People working with yeast and bacteria deliberately screen for secreting mutants.

Fowler: You have to be careful here, as there is a key difference between microbial cells and plant cells. Many microbial cells actively excrete products; plant cells, in contrast, possess a large internal vacuole into which they secrete many of the products in which we are interested. I am not sure, therefore, that plants possess a secretion mechanism comparable to that of microbial systems.

Scowcroft: Yes, but quite a lot of research is being done on *B. subtilis* to enhance the rate of secretion. Why can't you find a plant cell that does secrete?

Fowler: I know of people who are trying to clone in the yeast genes for 'leakiness' and components of transporter systems.

Zenk: I think it's very important, Professor Fowler, that you are re-exploring the optimum configuration for the fermentors. I am glad that you now stress that the air-lift fermentor may be not the best solution for all these systems and that you are coming back to the stirred-tank designs. You mentioned that the CO_2 concentration may be one of the problems—that the metabolic CO_2 is washed out of these cells and this may be the cause for some of the reduction in secondary product yield. We should also consider ethylene, which is a plant hormone, which is definitely washed out. I wouldn't be surprised to find that a few other gaseous compounds are present in these fermentations, which would in one way or the other influence the system.

Fowler: The reason I have no data on the ethylene content is that we couldn't afford to purchase an ethylene detector for our mass spectrometer. I do agree with you, however, that ethylene concentrations could also play an important role in large-scale cell culture performance.

Zenk: As to the hairy root system, do you think that it would be possible to grow hairy roots on a commercial scale in stirred-tank fermentors or in air-lift fermentors?

Fowler: At the moment we are trying to grow a hairy root culture in a stirred-tank reactor. The first impression that we have is that the roots will be smashed up and thus we will be back with a cell suspension culture.

Rhodes: Simple stirred-tank reactors are inappropriate for growing transformed roots to high densities. Our early attempts to grow transformed *Nicotiana* roots in a stirred tank led to callus formation and ultimately we developed a transformed suspension culture. This was associated with a substantial loss of nicotine production.

Zenk: Would it be feasible to grow root primordia, which are smaller?

Fowler: We have had under test for some time a simple impellar system with a very low shear characteristic. It consists essentially of a large glass ball at the end of a long shaft, which is itself attached to the centre of the vessel head plate through a flexible coupling. The glass ball contains a large magnet and follows a circular path around the vessel almost mid-way between the centre of the vessel and the vessel wall. The circular motion is maintained by a magnetic coupling to a motor mounted underneath the vessel. The system has been particularly useful for cultures that are lumpy. I imagine that it would be useful for root primordia cultures. At the moment we have only used the system for up to five litres culture volume; I am not sure that the current design would be easy to scale up.

Hall: You mentioned doubling times for *Phaseolus vulgaris* of 65 hours. Do people use *P. vulgaris* cultures for anything?

Fowler: I believe that there is some work on lectins being done with these cultures.

Scowcroft: Coming back to the last part on the socio-legal, you might add scientific, aspects. Plants synthesize compounds in one organ and transport it to another organ. What do the regulatory people say about that? For example, part of the flavour of strawberries is made in the growing point and transported to the fruit. How does this fit in to the definition of what is a natural product?

Fowler: We raised this very point with the UK authorities over two years ago and are still awaiting an answer. I am becoming increasingly concerned that until we have some definitive guidance on points such as this we will find sponsors more and more reluctant to commit research funds to projects where not only the science but also the regulatory conditions relating to possible products could be seen as risks.

Zenk: This situation with vanilla is ridiculous because the bulk of vanillin is produced by hydrolysis and oxidation of wood, with all the endogenous material, resins and so on, around it. How can the FDA accept vanillin that is produced from wood but not vanillin from vanilla cell cultures?

Galun: I have a partial answer. In Europe there is a requirement that all dairy products have to contain natural vanilla. No one observes this requirement, because if you calculate how much natural vanilla comes into Europe, it's about 1% or less of what is claimed to be in dairy products. So there is 99% cheating. There are three grades: natural vanilla, artificial vanilla and natural-like vanilla This third type they may allow by fermentation but you will not be able to call a fermentation product natural—you will be able to call it natural-like or natural-extracted. But this is not good enough for European dairy products, because they require exactly natural, although everyone is cheating.

Fowler: In the present socio-political climate, there is a consumer-led movement, followed sheep-like by the regulatory authorities, to regard everything 'natural' as wonderful and everything 'synthetic' as bad. In Europe we have the classic case of E number classifications, where in order to define food additives and control their use, most have been given a specific number or code. Unfortunately, many housewives regard an E number as indicating that the material is synthetic and must have something wrong with it, otherwise why would it be given a number? That citric acid derived from *Aspergillus* or citrus fruit has an

E number shows how stupid the situation is, and provides some indication of the perception problem that we scientists face, particularly in Europe.

Galun: But if you tell the housewife that it is natural vanilla when it is actually chemically produced, is that OK?

General discussion

Riley: Chairman, a philosophical question has occurred to me in listening to the papers that have been given on the production of agricultural products from cells in culture. At the present phase of world history, we have an enormous surplus capacity to produce plant materials from agriculture. Is it not more sensible to produce plant products from plants rather than from cells in culture? On the one hand in the developed world we are seeking ways of keeping the agricultural industry active, but from the discussion here it seems that we are finding other ways of further diminishing demands on that industry.

Galun: Unfortunately, most of the exotic products cannot be produced in those countries that have agricultural surpluses. Vanilla is very difficult to grow, for instance, in Europe and the same may be true for others. On the other hand, I believe that the future of rural communities in general lies not in improving their agricultural productivity—in that way they will never achieve a really nice standard of living—but in the introduction of centres of industry and services. This would be the only solution because the efficiency of agriculture is such that prices are low. In principle, this is part of the argument of this symposium; I don't think that farmers should stay with only farming, otherwise they will never reach a high standard of living.

Riley: Yes, but of the examples that you raise, take vanilla. Despite what you say about cheating in Europe (p 252), vanilla is a principal export of, for example, Madagascar. Madagascar is desperately in need of earning foreign currency, why deprive it of that opportunity?

Galun: I don't know, people will never be willing to pay the price. The price difference between chemically produced vanilla and the natural product is about 500-fold, so companies will always cheat.

Withers: I am very sympathetic to Sir Ralph's point about the economies of developing countries. We are not talking simply about not elevating their economies, but about the risk of actually damaging them. Another example, comparable to vanilla, is the possibility of producing cocoa *in vitro*, which has been explored in the United States.

Barz: This is certainly a very complex question to which there are several answers. One point that is quite important for certain countries, is that phytopharmaceuticals sometimes have to be imported from countries which are politically unstable or unreliable. I think there is a certain interest in technologically developed countries to have a constant supply of important products.

Fowler: On this point, we recently had a discussion with the Ministry of Agriculture, Fisheries and Foods in the UK about a research programme concerned with imported food additives. One of the key features that would lead them to fund the programme was a possible reduction in the UK import bill.

Riley: I am surprised that during the meeting we haven't heard anything about the possibility of producing products not normally present in a plant. I've no doubt that Marc (van Montagu), given sufficient incentive, could integrate into tobacco the ability to produce human growth hormone. This would provide a way of using agriculture and the natural energy which is available to agriculture, rather than having to put energy into cell culture.

Zenk: This question of putting the growth hormone gene into tobacco has been carefully evaluated as to whether or not this would be economic on an industrial scale. The cost of purifying this growth hormone out of tens of thousands of compounds which are in the tobacco plant is enormous. It is much better to put the growth hormone production into a yeast cell, let it be excreted from the cell, then purify it from maybe only a hundred compounds which are present in the spent medium.

van Montagu: Some of the techniques that we are using now for identifying enzymes and proteins in plants avoid this problem of purification. We heard from Professor Yamada what hard work purification is and we molecular biologists often don't know how to handle enzymes. In our group what we do is to partially enrich for a particular compound and then go directly to 2-D protein gels. We then isolate the protein (in very small amounts), microsequence it, and once we have identified part of the sequence, then we can synthesize the peptide. From there you use the peptide to raise antibodies in a mouse and then you can check that it is really the enzyme that you were looking for. If the antibody inactivates your enzyme, then by that time you already have a lot of data and you are on the way to cloning the gene.

Yamada: The first step is still to get the enzyme and to purify it.

van Montagu: No, that's just what I wanted to point out—you don't have to purify the protein.

Zenk: But one first has to know the biosynthetic pathway and the catalytic function of the protein

van Montagu: I agree, but this type of work you have done.

Potrykus: The beauty of this is that the pathways and the key enzymes are already known.

van Montagu: You can do it with membrane-bound proteins as well, that are very hard to purify.

Potrykus: I like Sir Riley's question because this discussion has been going on for at least ten years. I remember quite a number of meetings where a similar question was asked and I was often the one who asked it. I personally am convinced that nothing is as cheap as a plant growing in the field. As far as I understand the situation of harvesting material from plants for pharmaceutical purposes, people are using wild plants. There must be an enormous potential for progress via normal breeding.

van Montagu: When Ingo (Potrykus) says that we have been asking these questions for ten years, we must realize that the concepts have been there for ten years but the technology has not been available for that long. The number of groups working on genetic engineering in plants is still rather limited. It is in the mind of many people to do this research but it turns out that not so many are sufficiently well prepared to do it. Throughout the world universities are having their budgets for fundamental research cut. The knowledge needed to enable us to do these commercially interesting gene transfers has to come from basic plant biochemistry. Unfortunately, industry only supports research when the final product is there, after it has been demonstrated that there is a commercial application. So we want to go on looking at the necessary fundamental issues, but we are hanging in limbo—we know that we can do it but we need the funds to enable us to do it.

Potrykus: You are right that the techniques were not available to answer specific questions, but I meant the general philosophy.

Scowcroft: In connection with that same issue, Sir Ralph, one of the things that I have been exposed to over the last twelve months is that, in addition to conducting research to enhance productivity, there is an increasing emphasis, with particular influence from the food industry, on research to enhance consumer perceived preferences in foods, i.e. flavour, texture, colour and shelf-life. The food industry sector is now saying to people in our business, give us a range of varieties of, say, oil seed rape, with different fatty acid compositions, such as high oleic acid or high stearic acid. That saves the food industry production costs in their blending operations. This is one of the directions in which I think agriculture is heading. The task for us is to come up with high quality, specialized types of crop varieties.

Potrykus: Something that I found exciting during this last day was that there are so many key enzymes known for important metabolic pathways, one can see that it will not take many years to isolate the genes. If we had these genes, I would put them into maize or tobacco and ask these plants to produce these compounds, instead of asking a cell culture. There is so much knowledge now available of the enzymology of secondary metabolite production, that it's surprising that, as yet, there is no collaboration between molecular biologists and laboratories studying secondary product synthesis. I think that a few people trained in molecular biology, working in a lab such as Professor Yamada's or Professor Zenk's would soon isolate a number of commercially exploitable genes from this material.

Hall: Within this discussion of the various products was the question of putting pharmaceuticals and so on into plants. What I perhaps failed to bring out strongly in my own paper, was that the purpose of starting to look for the localization of products encoded by the introduced genes within cells relates to the ability to do these ideas. As Marc (van Montagu) says, several of the ideas have been around for a long time but one simply does not have the stability in plant cells of any of these products. This is a real problem. I mentioned at the end of my paper the *Bacillus thuringiensis* (BT) project to develop insect resistance in plants, which I think is widely seen as a very valuable bioengineering project. It is valuable in many ways, including the total productivity and the sociological considerations that must be taken into account. However, there is a major problem, namely that the BT protein is extremely unstable and no one has actually seen that protein produced in the plants even though the plants do have biological activity. So I think that many of the people working directly in the area of bioengineering are recognizing that they have got to go back and look at these very fundamental problems. Then we will be able to develop many of these attractive esoteric ideas of enhancing specific value-added products, engineering seed proteins so that they have the type of processing characteristics that Dr Scowcroft mentioned, and so on.

van Montagu: You say that nobody has seen the intact BT protein, we have seen it but the amount is extremely low and it is degraded. But you don't need the intact protein, if you take the first 607 amino acids, all the activity is there, and the amount of that fragment present is quite high.

Hall: Nevertheless, Marc, I think you would agree that the amount of protein there is clearly extremely low. When you look at the promoter in front of that protein coding sequence, if you put another coding sequence behind that promoter, you would see the protein. It is clear that the BT protein is not very stable in the plant cell.

van Montagu: You mean even the 607 amino acid fragment?

Hall: If you put the coding sequence for a stable protein behind that promoter, you would see a band for that protein within a total protein preparation from that cell. Since you do not see that protein, it seems clear that the BT peptide is not stable.

Harms: For practical purposes, it doesn't matter whether the protein is stable as long as it kills the insect and that has been demonstrated.

Hall: But it doesn't kill it as effectively as you would like. If it were more stable, it would kill more rapidly.

Potrykus: To return to Sir Riley's second question: the experiment I would like to see done in the near future is that as soon as the gene for one of the interesting secondary metabolites has been isolated, this gene is transferred into higher plants, into yeast, into bacteria and into plant cell cultures. A potential which has not been used is to increase the productivity of cell cultures by linking these genes to strong constitutive promoters. Then we should compare the production and the costs of this product in the field, in the fermenter, in yeast and in bacteria. Only on the basis of such data could we continue this discussion.

Scowcroft: The other point in support of what Sir Ralph was saying is that we tend to think of plants producing these compounds on a continuous basis. In this case, extraction and processing plants will require a continuous input of fresh plant material. Plant seeds are a remarkably good way to store plant products. Seeds can be stored over a long period of time and thus maintain a constant supply of feedstock to operating plants. If we begin to think about using plants as alternative systems to produce some of these compounds in seeds, this would provide a tremendous advantage from a storage standpoint over production in the leaf, for example, where continuous harvesting is required.

Potrykus: There are other storage organs in plants, for example tubers.

Scowcroft: Tubers are not suitable for long-term storage.

Galun: It makes a lot of sense to grow those natural products in developing countries, but it would be even better if the affiliated industry would move to those countries. I don't see any reason why Jamaica couldn't be a centre of cacao production. Why do they have to export all their cacao? I said the same recently in Bogota, all the production based on coffee could actually stay with them, why should it be done thousands of miles away? The same is true of pharmaceutical products: the difference in price between the raw material and the final product is sometimes enormous. We all know that the price of corn in a cornflakes box is about 1% of the price of the box. It's alarming but it is true. If all the processing and packaging could be done in developing countries, it would be much better. The main effort should be to bring some industrialization to those countries and to let them have the bigger share of the final product price. What we are doing is perpetuating the situation—you produce the raw material and in a better quantity and then ship everything back to Europe or the United States, there the companies will make the profit—I think this is the wrong conclusion.

Index of contributors

Non-participating co-authors are indicated by asterisks. Entries in bold type indicate papers; other entries refer to discussion contributions

Indexes compiled by John Rivers

Subject index

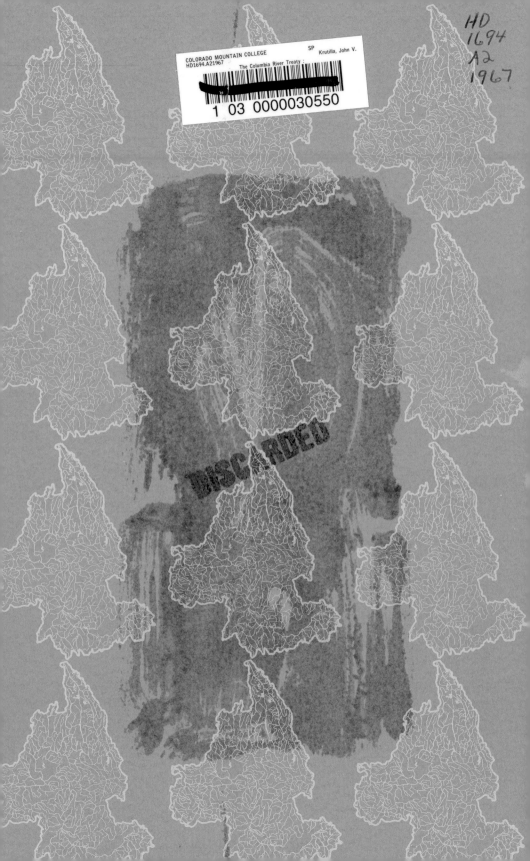

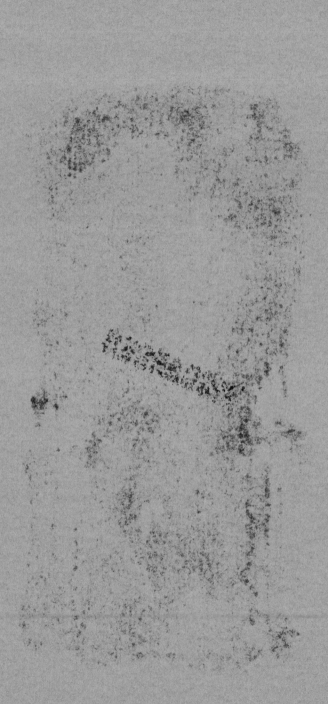